ISBN 978-1-334-33969-1
PIBN 10564951

RESEARCHES,

[CH]EMICAL AND PHILOSOPHICAL;

CHIEFLY CONCERNING

NITROUS OXIDE,

OR

PHLOGISTICATED NITROUS AIR,

AND ITS

RESPIRATION.

By HUMPHRY DAVY,

[SUP]ERINTENDENT OF THE MEDICAL PNEUMATIC
INSTITUTION.

LONDON:

CONTENTS.

DIVISION II.

DIVISION III.

DIVISION IV.

EXPERIMENTS and OBSERVATIONS on the compofition of
NITROUS GAS, and on its abforption by different bodies.

DIVISION V.

EXPERIMENTS and OBSERVATIONS on the production of
NITROUS OXIDE from NITROUS GAS and NITRIC
ACID in different modes.

RESEARCH II.

Into the combinations of NITROUS OXIDE, *and its decompofition.*

DIVISION I.

EXPERIMENTS and OBSERVATIONS on the combinations
of NITROUS OXIDE.

DIVISION II.

Decompofition of NITROUS OXIDE by combuftible Bodies.

RESEARCH III.

Relating to the RESPIRATION *of* NITROUS OXIDE *and* OTHER GASES.

1 2

DIVISION I.

EXPERIMENTS and OBSERVATIONS on the effects produced upon Animals by the respiration of NITROUS OXIDE.

DIVISION II.

Of the changes effected in NITROUS OXIDE and other Gases, by the Respiration of Animals.

RESEARCH IV.

Relating to the EFFECTS *produced by the* RESPIRATION *of*
NITROUS OXIDE *upon different* INDIVIDUALS.

DIVISION I.

HISTORY of the Difcovery.—EFFECTS produced by the
Refpiration of different GASES.

DIVISION II.

DETAILS of the Effects produced by the Refpiration of
NITROUS OXIDE upon different Individuals, furnifhed
by Themfelves.

DIVISION III.

Abſtracts from additional Details—Obſervations on the effects of NITROUS OXIDE, by Dr. BEDDOES—Concluſion.

APPENDIX.

INTRODUCTION.

In confequence of the discovery of the respirability and extraordinary effects of nitrous oxide, or the dephlogifticated nitrous gas of Dr. Prieftley, made in April 1799, in a manner to be particularly defcribed hereafter, * I was induced to carry on the following inveftigation concerning its compofition, properties, combinations, and mode of operation on living beings.

In the courfe of this inveftigation, I have met with many difficulties; fome arifing from the novel and obfcure nature of the fubject, and

* A fhort account of this difcovery has been given in Dr. Beddoes's Notice of fome Obfervations made at the Pneumatic Inftitution, and in Mr. Nicholfon's Phil. Journal for May and December 1799.

others from a want of coincidence in the observations of different experimentalists on the properties and mode of production of the gas. By extending my researches to the different substances connected with nitrous oxide; nitrous acid, nitrous gas and ammoniac ; and by multiplying the comparisons of facts, I have succeeded in removing the greater number. of those difficulties, and have been enabled to give a tolerably clear history of the combinations of oxygene and nitrogene.

By employing both analysis and synthesis whenever these methods were equally applicable, and comparing experiments made under different circumstances, I have endeavoured to guard against sources of error; but I cannot flatter myself that I have altogether avoided them. The physical sciences are almost wholly dependant on the minute observation and comparison of properties of things not immediately obvious to the senses: and from the difficulty of discovering every possible mode of examination, and from the modification of per ceptions by the ftate of feeling, it appears nearly impoffible that all the relations of a feries of phænomena can be difcovered by a fingle inveftigation, particularly when thefe relations are complicated, and many of the agents unknown. Fortunately for the active and progreffive nature of the human mind, even experimental refearch is only a method of approximation to truth.

In the arrangement of facts, I have been guided as much as poffible by obvious and fimple analogies only. Hence I have feldom entered into theoretical difcuffions, particularly concerning light, heat, and other agents, which are known only by ifolated effects.

Early experience has taught me the folly of hafty generalifation. We are ignorant of the laws of corpufcular motion ; and an immenfe mafs of minute obfervations concerning the more complicated chemical changes muft be collected, probably before we fhall be able to afcertain even whether we are capable of difcovering them. Chemiftry in its prefent ftate, is fimply a partial hiftory of phænomena, confifting of many feries more or lefs extenfive of accurately connected facts.

With the moft important of thefe feries, the arrangement of the combinations of oxygene or the antiphlogiftic theory difcovered by Lavoifier, the chemical details in this work are capable of being connected.

In the prefent ftate of fcience, it will be unneceffary to enter into difcuffions concerning the importance of inyeftigations relating to the properties of phyfiological agents, and the changes effected in them during their operation. By means of fuch inveftigations, we arrive nearer towards that point from which we fhall be able to view what is within the reach of difcovery, and what muft for ever remain unknown to us, in the phænomena of organic life. They are of immediate utility, by enabling us to extend our analogies fo as to inveftigate the properties of untried fubftances, with greater accuracy and probability of fuccefs.

The firſt Reſearch in this work chiefly relates to the production of nitrous oxide and the analyſis of nitrous gas and nitrous acid. In this there is little that can be properly called mine; and if by repeating the experiments of other chemiſts, I have ſometimes been able to make more minute obſervations concerning phænomena, and to draw different concluſions, it is wholly owing to the uſe I have made of the inſtruments of inveſtigation diſcovered by the illuſtrious fathers of chemical philoſophy,* and ſo ſucceſsfully applied by them to the diſcovery of truth.

In the ſecond Reſearch the combinations and compoſition of nitrous oxide are inveſtigated, and an account given of its decompoſition by moſt of the combuſtible bodies.

The third Reſearch contains obſervations on the action of nitrous oxide upon animals, and

* Cavendiſh, Prieſtley, Black, Lavoiſier, Scheele, Kirwan, Guyton, Berthollet, &c.

an investigation of the changes effected in it by respiration.

In the fourth Research the history of the respirability and extraordinary effects of nitrous oxide is given, with details of experiments on its powers made by different individuals.

I cannot close this introduction, without acknowledging my obligations to Dr. Beddoes. In the conception of many of the following experiments, I have been aided by his conversation and advice. They were executed in an Institution which owes its existence to his benevolent and philosophic exertions.

Dowry-Square, Hotwells, Bristol.
June 25th, 1800.

RESEARCH I.

CONCERNING THE ANALYSIS

OF

NITRIC ACID AND NITROUS GAS.

AND

THE PRODUCTION OF

NITROUS OXIDE.

RESEARCH I.

INTO THE PRODUCTION AND ANALYSIS

OF

NITROUS OXIDE,

AND

THE AERIFORM FLUIDS RELATED TO IT.

DIVISION I.

EXPERIMENTS and OBSERVATIONS on the composition of NITRIC ACID, *and on its combinations with* WATER *and* NITROUS GAS.

I. THOUGH fince the commencement of Pneumatic Chemiftry, no fubftance has been more the fubject of experiment than Nitrous Acid; yet ftill the greateft uncertainty exifts with regard to the quantities of the principles entering into its compofition.

In comparing the experiments of the illustrious Cavendifh on the fynthefis of nitrous acid, with thofe of Lavoifier on the decompofition of nitre by charcoal, we find a much greater difference in the refults than can be

A

accounted for by suppofing the acid formed, and that decompofed, of different degrees of oxygenation.

In the moft accurate experiment of Cavendifh, when the nitrous acid appeared to be in a ftate of deoxygenation, 1 of nitrogene combined with about 2,346 of oxygene.* In an earlier experiment, when the acid was probably fully oxygenated, the nitrogene employed was to the oxygene nearly as 1 to 2,92.†

Lavoifier, from his experiments on the decompofition of nitre, and combination of nitrous gas and oxygene, concludes, that the perfectly oxygenated, or what he calls nitric acid, is compofed of nearly 1 nitrogene, with 3,9 of oxygene ; and the acid in the laft ftate of deoxygenation, or nitrous acid, of about 3 oxygene with 1 nitrogene.‡

---

* Phil. Tranf. v. 78, p. 270.   † Phil. Tranf. v. 75, p. 381,

‡ Elem. Kerr's Tranf. page 76, and 216, and Mem, des Sav. Etrang. tom. 7, page 629.

ₜ Great as the difference is between the estimations of thefe philofophers, we find differences ftill greater in the accounts of the quantities of nitrous gas neceffary to faturate a given quantity of oxygene, as laid down by very accurate experimentalifts.' On the one hand, Prieftley found 1 of oxygene condenfed by 2 of nitrous gas, and Lavoifier by $1\frac{7}{8}$. On the other, Ingenhouz, Scherer, and De la Metherie, ftate the quantity neceffary to be from 3. to 5:* Humbolt, who has lately inveftigated Eudiometry with great ingenuity, confiders the mean quantity of nitrous gas neceffary to faturate 1ₜ of oxygene, as about 2,55.†

II. To reconcile thefe different refults is impoffible, and the immediate connection of the fubject with the production of nitrous oxide, as well as its general importance, obliged me to fearch for means of accurately deter-

* Ingenhouz fur les Vegetaux, pag. 205. De la Metherie. Effai fur differens Airs, pag. 252.

† Annales de Chimie, tome 28, p. 168.

mining the composition of nitrous acid in its different degrees of oxygenation.

The first defideratum was to afcertain the nature and compofition of a fluid acid, which by being deprived of, or combined with nitrous gas, might become a ftandard of comparifon for all other acids.

To obtain this acid I fhould have preferred the immediate combination of oxygene and nitrogene over water by the electric fpark, had it been poffible to obtain in this way by a common apparatus fufficient for extenfive examination; but on carefully perufing the laborious experiments of Cavendifh, I gave up all thoughts of attempting it.

My firft experiments were made on the decompofition of nitre, formed from a known quantity of pale nitrous acid of known fpecific gravity, by phofphorus, tin, and charcoal : but in thofe proceffes, unafcertainable quantities of nitrous acid, with excefs of nitrous gas, always efcaped undecompounded, and from the non-coincidence of refults, where different quanti-

ties of combuftible fubftances were employed,
I had reafons for believing that water was
generally decompofed.

Before thefe experiments were attempted, I had
analized nitrous gas and nitrous oxide, in a manner to be particularly defcribed hereafter; fo
that a knowledge of the quantities of nitrous
gas and oxygene entering into the compofition
of any acid, enabled me to determine the proportions of nitrogene and oxygene it contained.
In confequence of which I attempted to combine together oxygene and nitrous gas, in fuch
a manner as to abforb the nitrous acid formed
by water, in an apparatus by which the quantities of the gafes employed, and the increafe of
weight of the water, might be afcertained; but
this procefs likewife failed. It was impoffible
to procure the gafes perfectly free from nitrogene, and during their combination, this nitrogene made to pafs into a pneumatic apparatus
communicating with a veffel containing the
water carried over with it, much nitrous acid
vapor, of different compofition from the acid
abforbed.

After many unfuccefsful trials, Dr. Prieftley's experiments on nitrous vapor * induced me to fuppofe that oxygene and nitrous gas, made to combine out of the contact of bodies having affinity for oxygene, would remain permanently aëriform, and on throwing them feparately into an exhaufted glafs balloon, I found that this was actually the cafe ; increafe of temperature was produced, and orange colored nitrous acid gas formed, which after remaining for many days in the globe, at a temperature below 56°, did not in the flighteft degree condenfe.

This fact afforded me the means not only of forming a ftandard acid, but likewife of afcertaining the fpecific gravity of nitrous acid in its aëriform ftate.

III. Previous to the experiment, for the purpofe of correcting incidental errors, I was induced to afcertain the fpecific gravity of the

* Experiments and Obfervations, Vol. iii. laft edition, page 105, &c.

gafes employed, particularly as I was unacquainted with any procefs by which the
weight of nitrous gas had been accurately
determined. Mr. Kirwan's eftimation, which
is generally adopted, being founded upon the
comparifon of the lofs of weight of a folution
of copper in dilute nitrous acid, with the quantity of gas produced.*

The inftruments that I made ufe of for containing and meafuring my gafes, were two mercurial airholders graduated to the cubic inch of
Everard, and furnifhed with ftop cocks.†

* When copper is diffolved in dilute nitrous acid, certain
quantities of nitrogene are generally produced, likewife the
nitrous gas carries off in folution fome nitrous acid.

† This airholder, confidered as a pneumatic inftrument,
is of greater importance, and capable of a more extenfive
application than any other. It was invented by Mr. W.
CLAYFIELD, and in its form is analogous to Mr. WATT's
hydraulic bellows, confifting of a glafs bell playing under
the preffure of the atmofphere, in a fpace between two cylinders filled with mercury. A particular account of it will
be given in the appendix.

They were weighed in a glafs globe, of the capacity of 108 cubic inches, which with the fmall glafs ftop-cock affixed to it, was equal, when filled with atmofpheric air, to 1755 grains. The balance that I employed, when loaded with a pound, turned with lefs than one eighth of a grain.

Into a mercurial airholder, of the capacity of 200 cubic inches, 160 cubic inches of nitrous gas were thrown from a folution of mercury in nitrous acid.

70 meafures of this were agitated for fome minutes in a folution of fulphate of iron,* till the diminution was complete. The nitrogene remaining hardly filled a meafure.; and if we fuppofe with Humbolt † that a very fmall portion of it was abforbed with the nitrous gas, the whole quantity it contained may be, eftimated at 0,0142, or $\frac{1}{70}$.

75 cubic inches received from the airholder

---

* This abforption will be hereafter particularly treated of.

† Annales de Chimie. Tome xviii. page 139.

into an exhaufted balloon, increafed it in weight
25,5 grains ; thermometer being . 56°, and
barometer 30,9. And allowing for the fmall
quantity of nitrogene in. the gas, 100 cubic
inches of it will weigh 34.3 grains..

One hundred and. thirty cubic inches of
oxygene were procured from oxide of manganefe and fulphuric acid, by heat, and received in
another mercurial airholder. .

10 . meafures .of it, mingled with 26 of the
nitrous gas,. gave, after the refiduum was expofed to folution of fulphate of iron, rather
more than one meafure. Hence we may conclude that it contained about 0,1 nitrogene.

60 cubic inches of it weighed 20,75 grains ;
and accounting for the, nitrogene contained in
thefe, 100 grains of pure oxygene will weigh
35,09 grains.

Atmofpherical air was decompofed by nitrous
gas in excefs; and the refiduum wafhed with
folution of fulphate of iron till the Nitrogene
remained pure ; 87 cubic inches of it weighed
26,5 grains, thermometer being 48°, barometer
30,1 ; 100 will confequently weigh 30,45.

90 cubic inches ·of the air of· the laboratory not deprived of· its ·carbonic· acid, weighed 28,75 grains ; thermometer 53, barometer 30 : 100 cubic inches will confequently ·weigh 31,9.* 16 meafures of this air, with 16 ·nitrous ·gas, of known ·compofition, · diminifhed to 19. Hence it contained about ;26 oxygene.§

In comparing my refults with thofe of Lavoifier and Kirwan, the eftimation of the weights of ni- trogene and oxygene is very little different, the corrections for temperature ·and ·preffure· being made, from that of thofe celebrated philofophers. The firft makes oxygene to weigh ☦ 34,21, and nitrogene 30,064 per cent ; and the laft, oxy- gene 34, ‡ and nitrogene 30,5.

---

* A table of the fpecific gravities of thefe gafes, and other gafes, hereafter to be mentioned, reduced to a barometrical and thermometrical ftandard, will be given in the appendix.

§ 40 meafures, expofed to folution of potafh, gave an abforption of not quite a quarter of a meafure : hence it contained an inconfiderable quantity of carbonic acid.

† Traité Elementaire.

‡ Effai fur le phlogiftique, page 30.

The fpecific gravity of nitrous gas, according to Kirwan, is to that of common air as 1194 to 1000. Hence it fhould weigh about 37 grains per cent. This difference from my eftimation is not nearly fo great as I expected to have found it.§

IV.* The thermometer in the laboratory ftanding at 55°, and the barometer at 30,1, I now proceeded to my experiment. The oxygene that I employed was of the fame compofition as that which I had previoufly weighed. The nitrous gas contained ,0166 nitrogene.

For the purpofe of combining the gafes, a glafs balloon was procured, of the capacity of 148 cubic inches, with a glafs ftop-cock adapted to it, having its upper orifice tubulated and graduated for the purpofe of containing and meafuring a fluid. The whole weight of this globe and its appendages, when filled with common air, was 2066,5 grains.

§ The diminution of the fpecific gravity of the gas from the quantity of nitrogene evolved in his experiment, probably deftroyed, in fome meafure, the fource of error from the nitrous acid carried over.

* Experiment I.

It was partially exhausted by the air-pump, and lost in weight just 32 grains. From whence we may conclude that about 15 grains of air remained in it.

In this state of exhaustion it was immediately cemented to the stop-cock of the mercurial airholder, and the communication being made with great caution, 82 cubic inches of nitrous gas rushed into the globe, on the outside of which a slight increase of temperature was perceived, while the gases on the inside appeared of a deep orange.

Before the common temperature was restored, the communication was stopped, and the globe removed. The increase of weight was 29,25 grains; whence it appeared that 1,14 grains of common air, part of which had been contained in the stop-cocks, had entered with the nitrous gas.

Whilst it was cooling, from the accidental loosening of the stopper of the cock, 3 grains more of common air entered.*

---

* That no greater contraction took place depended on the solution of the nitrous acid formed in the nitrous gas; a phænomenon to be explained hereafter.

The communication was now made between the globe and the mercurial airholder containing oxygene. 64 cubic inches were flowly preffed in, when the outfide of the globe became warmer, and the color on the infide changed to a very dark orange. As it cooled, 6 cubic inches more flowly entered; but no new increafe of temperature, or change of color took place.

The globe being now completely cold, was flopped, removed, and weighed; it had gained 24,5 grains, from whence it appears that 0,4 grains of common air contained in the flopcocks, had entered with the oxygene.*

To abforb the nitrous acid gas, 41 grains of water were introduced by the tube of the flopcock, which though clofed as rapidly as poffible, muft have fuffered nearly ,5 grains of air to

* I judged it expedient always to afcertain the quantity of air in the flop-cocks by weight, as it was impoffible to join them fo as to have always an equal capacity. The upper tubes of the two flop-cocks not joined, contained nearly an inch and half.

enter at the same time, as the increafe of weight was 41,5 grains. The dark orange of the globe diminifhed rapidly; it became warm at the bottom, and moift on the fides. After a few minutes the color had almoft wholly difappeared.

To afcertain the quantity of aëriform fluid abforbed, the globe was again attached to the mercurial air apparatus, containing 140 cubic inches of common air. When the communication was made, 51 cubic inches rufhed in, and it gained in weight 16,5 grains.

A quantity of fluid equal to 54 grains was now taken out of the globe. On examination it proved to be flightly tinged with green, and occupied a fpace equal to that filled by 41,5 grains of water. Its fpecific gravity was confequently 1,301.

To afcertain if any unabforbed aëriform nitrous acid remained in the globe, 13 grains of folution of ammonia were introduced in the fame manner as the water, and after fome minutes, when the white vapor had condenfed, the communication was again made with the mercurial airholder containing common air. A minute quantity entered, which could not be eſtimated at more than three fourths of an inch, and the globe was increaſed in weight about 13,25 grains.*

Common air was now thrown into the globe till the reſidual gaſes of the experiment were judged to be diſplaced; it weighed 2106,5 grains, that is, 40 grains more than it had weighed when filled with common air before the experiment.†

* That is, by the ſolution of ammonia, and air.

† The following is an account of the increaſe and dimi‑ nution of weight of the globe, as it was noted in the jour‑ nal.

Globe filled with common air   -   gr. 2066,5
After exhauſtion   -   -   -   2034,5
After introduction of nitrous gas, 82
     cubic inches   -   -   2004,25
After the accidental admiſſion of com‑
     mon air   -   -   -   2007,25
After the admiſſion of oxygene   -   2091,75
——— / ——— 41 grains of water   2133,25
——— ——— 51 cubic inches of air   2149,75
Taken out 54 grains of ſolution   -   2095,75
Introduced 13 grains of ammoniacal
     ſolution   -   -   -   2109,25
After introduction of common air   -   2106,5

· ' And if from thofe 40 grains we take 13 for the folution of ammonia introduced, the re-mainder, 27, will be the quantity of folution of nitrous acid in water remaining in the globe, which added to 54, equals 81 grains, the whole quantity formed ; but if from this be taken 41; grains, the quantity of water, the remainder 40 grains; will be the quantity of nitrous acid gas abforbed in the folution.

To find the abfolute quantity of nitrous acid formed, we muft find the fpecific gravity of that abforbed ; but as during, and after its abforption, 17 grains of air, equal to 53,2 cubic inches entered, it evidently filled fuch a fpace. 53,2 cubic inches of it confe-quently weigh 40 grains, and 100 cubic inches 75,17 grains. Then ,75 cubic inches weigh ,56 grains, and this added to 40, makes 40,56 grains, equal to 57,0 cubic inches, the whole quantity of aëriform nitrous acid pro-duced.

But the quantity of nitrous gas entering into this, allowing for the nitrogene it contained, is

27,6 grains, equal to about 80,5 cubic inches; and the oxygene is 40,56 — 27,6 = to 12,96 grains, or 36,9 cubic inches.

V. There could exist in this experiment no circumstance connected with inaccuracy, except the impoffibility of very minutely determining the quantities of common air which entered with the gafes from the ftop-cocks. But if errors have arifen from this fource, they muft be very inconfiderable; as will appear from a calculation of the fpecific gravity of the nitrous acid gas, founded on the volume of the gafes that entered the globe.

The air that remained in the globe
after exhauftion was 15 grains = 47* cub. in.
The nitrous gas introduced was · 82
Common air   -   -   -   13
Oxygene   -   -   -   -   70
Common air·   -   -   -   1
                                   ———

* Decimals are omitted, becaufe the excefs of the two firft numbers is exactly corrected by the deficiency of the laft.

Whole quantity of air thrown into
 the globe   213
From which fubtract its capacity   148
            ———
The remainder is   65

And this remainder taken from 80,5 nitrous
gas + 36,9 oxygene, leaves, 52,4 cubic inches,
which is the fpace occupied by the nitrous acid
gas, and which differs from 53,95 only by 1,55
cubic inches.

 I ought to have obferved, that before this
conclufive experiment, two fimilar ones had
been made. In comparing the refults of one
of them, performed with the affiftance of my
friend, Mr. JOSEPH PRIESTLEY, Dr. PRIEST-
LEY's eldeft fon, and chiefly detailed by him
in the journal, I find a coincidence greater than
could be even well expected, where the proceffes are fo complex. According to that
experiment, 41,5 grains of nitrous acid gas
fill a fpace equal to 53 cubic inches, and
are compofed of nearly 29 nitrous gas, and
12,5 oxygene.

We may then conclude, First, that 100 cubic inches of nitrous acid, such as exists in the * aëriform state saturated with oxygene, at temperature 55°, and atmospheric pressure 30,1 weigh 75,17 grains.

Secondly, that 100 grains of it are composed of 68,06 nitrous gas, and 31,94 oxygene. Or assuming what will be hereafter proved, that 100 parts of nitrous gas consist of 55,95 oxygene, and 44,05 nitrogene, of 29,9 nitrogene, and 70,1 oxygene ; or taking away decimals, of 30 of the one to 70 of the other.

Thirdly, that 100 grains of pale green solution of nitrous acid in water, of specific gravity 1,301, is composed of 50,62 water, and 49,38 acid of the above composition.

VI. Having thus ascertained the composition of a standard acid, my next object was to obtain it in a more condensed state, as it was otherwise impossible to saturate it to its full

---

* As is evident from the superabundant quantity of oxygene thrown into the giobe.

extent with nitrous gas. But this I could effect in no other way than by comparing mixtures of known quantities of water, and acids of different fpecific gravities and colors, with the acid of 1,301.

For the purpofe of combining my acids with water, I made ufe of a cylinder about 8 inches long, and ,3 inches in diameter, accurately graduated to grain meafures, and furnifhed with a very tight ftopper.

The concentrated acid was firft flowly poured into it, and the water gradually added till the required fpecific gravity was produced;* the cylinder being clofed and agitated after each addition, fo as to produce combination without any liberation of elaftic fluid.

After making a number of experiments with

---

* The weight of the acid poured into the cylinder being known, its fpecific gravity was known from the fpace it occupied in the phial. The weight of water being likewife known, the fpecific gravity of the folution, when the common temperature was produced, was given by the condenfation.

acids of different colors in this advantageous way, I at length found that 90 grains of a deep yellow acid, of fpecific gravity 1,5, became, when mingled at 40° with 77,5 grains of water, of fpecific gravity 1,302, and of a light green tinge, as nearly as poffible refembling that of the ftandard acid.

Suppofing, then, that thefe acids contain nearly the fame relative proportions of oxygene and nitrogene, 100 grains of the deep yellow acid of 1,5, are compofed of 91,9 grains true nitrous acid,† and 8,1 grains of water.

To afcertain the difference between the compofition of this acid, and that of the pale, or nitric acid, of the fame fpecific gravity, I inferted 150 grains of it into a fmall cylindrical mattrafs of the capacity of ,5 cubic inches, accuratcly graduated to grain meafures, and connected by a curved.

---

† That is, fuch as it exifts in the aëriform ftate at 55o. From the ftrong affinity of nitrous acid for water, we may1 fuppofe that this acid gas contains a larger proportion of it than the other gafes.

tube with the water apparatus. After heat had been applied to the bottom of the mattrafs for a few minutes, the color of the fluid gradually changed to a deep red, whilft the globules of gas formed at the bottom of the acid, were almoft wholly abforbed in paffing through it. In a fhort time deep red vapour began to fill the tube, and being condenfed by the water in the apparatus, was converted into a bright green fluid, at the fame time that minute globules of gas were given out. As the heat applied became more intenfe, a very fingular phænomenon prefented itfelf; the condenfed vapor, increafed in quantity, at length filled the curvature of the tube, and when expelled, formed itfelf into dark green fpherules, which funk to the bottom of the water, refted for a moment, and then refolved themfelves into nitrous gas.*

When the acid was become completely pale, it was fuffered to cool, and weighed. It had loft near 15 grains, and was of fpecific gravity

* This appearance will be explained hereafter.

1,491. 2 cubic inches and quarter of nitrous gas only. were collected.

From this experiment evidently no conclufions could be drawn, as the nitrous gas had carried over with it much nitrous acid (in the form of what Dr. Prieftley calls nitrous vapor) and was partially diffolved with it in the water.†

To afcertain, then, the difference between the pale and yellow acids, I was obliged to make ufe of fynthefis, compared with analyfis, carried on in a different mode, by means of the following apparatus.

VII. To the ftop-cock of the upper cylinder of the mercurial airholder, a capillary tube was adapted, bent fo as to be capable of introduction into an orifice in the ftopper of a graduated phial fimilar to that employed for mingling acids with water, and fufficiently long to reach the bottom. With another orifice in the ftopper of the phial was connected a fimilar tube cur-

---

† This phænomenon will be particularly explained hereafter.

ved, for the purpofe of containing a fluid, and of increafed diameter at the extremity.*

50 cubic inches of pure nitrous gas † were thrown into the mercurial apparatus. The graduated phial, containing 90 grains of nitric acid, of fpecific gravity 1,5, was placed on the top of the airholding cylinder, and made to communicate with it by means of the ftop-cock and firft tube. Into the fecond tube a fmall quantity of folution of potafh was placed. When all the junctures were carefully cemented, by preffing on the air-holder, the nitrous gas was flowly paffed into the phial, and abforbed by the nitrous acid it contained; whilft the fmall quantities of nitrogene evolved, flowly drove forward the folution in the curved tube; from the height of which, as compared with that of the mercury in the conducting tube, the preffure on the air in the cylinder was known.

* The outline only of this apparatus is given here, as far as was neceffary to make the experiment intelligible; a detailed account of it, and of its geneial application, will be given in the appendix.

† That is, from nitrous acid and mercury.

In proportion as the nitrous gas was abforbed, the phial became warm, and the acid changed color; it firft became ftraw-colored, then pale yellow, and when about $7\frac{1}{2}$ cubic inches had been combined with it, bright yellow. It had gained in weight nearly 3 grains, and was become of fpecific gravity 1,496.

This experiment afforded me an approximation to the real difference between nitric and yellow nitrous acid ; and learning from it that nitric acid was diminifhed in fpecific gravity by combination with nitrous gas, I procured a pale acid of fpecific gravity 1,504.* After this acid had been combined in the fame manner as before, with about 8 cubic inches of nitrous gas,§ it became nearly of fpecific gravity 1,5, and had gained in weight about 3 grains.

* A pale acid of 1.52, by being converted into yellow acid, became nearly of fpecific gravity 15,1.

§ It is impoffible to afcertain the quantity of gas abforbed to more than a quarter of a cubic inch, as the firft portions of nitrous gas thrown into the graduated cylinder are combined with the oxygene of the common air in it, to form nitrous acid, and hence the flight excefs of weight.

Affuming the accuracy of this experiment as a foundation for calculation, I endeavoured in the fame manner to afcertain the differences in the compofition of the orange-colored acids, and the acids containing ftill larger proportions of nitrous gas.

93 grains of the bright yellow acid of 1,5 became, when 6 cubic inches of gas had been paffed through it, orange colored and fuming; whilft the undiffolved gas increafed in quantity fo much as to render it impoffible to confine it by the folution of potafh. When 9 cubic inches had paffed through, it became dark orange. It had gained in weight 2,75 grains, and was become of fpecific gravity 1,48 nearly. Hence it was evident that much nitrous gas had paffed through it undiffolved. 25 cubic inches more of nitrous gas were now flowly fent through it : it firft became of a light olive, then of a dark olive, then of a muddy green, then of a bright green, and laftly of a blue green. After its affumption of this color, the gas appeared to pafs through it unaltered, and large globules

of fluid, of a darker green than the reft, remained at the bottom of the cylinder, and when agitated, did not combine with it. The increafe of weight was only 1 grain, and the acid was of fpecific gravity 1,474 nearly.

In this experiment it was evident that the unabforbed nitrous gas had carried over with it a confiderable quantity of nitrous acid. I endeavoured to correct the errors refulting from this circumftance, by connecting the curved tube firft with a fmall water apparatus, and afterwards with a mercurial apparatus; but when the water apparatus was ufed, the greater part of the unabforbed gas was diffolved with the nitrous acid it held in folution, by the water; and when mercury was employed, the nitrous acid that came over was decompofed, and the quantity of nitrous gas evolved, in confequence increafed.

As it was poffible that a fmall deficiency of weight might arife from the red vapor given out during the proceffes of weighing and examining the acid in the laft experiment,

35 cubic inches of nitrous .gas were very
flowly paffed through 90 grains of pale. nitrous
acid, of.fpecific gravity 1,5 :. it. became of. fimilar appearance to that juft defcribed, had gained
in weight 6,75 grains, and was become of fpecific gravity .1,475.

Thefe experiments did not afford approximations fufficiently accurate towards the compofition of deoxygenated acids, containing more
nitrous gas than. the dark orange colored. To,
obtain. them, a folution confifting of 94,25
grains of blue green, or perfectly nitrated acid,
(if we may be allowed to employ the term), of
fpecific gravity 1,475, was inferted into a
graduated phial, and connected by a curved
tube, with the mercurial airholder; in the
conductor of which a fmall quantity of
water was inferted to abforb the nitrous acid
which might be carried over by the gas. Heat
was flowly applied to the phial, and nitrous
gas given out with great rapidity. When 4
cubic inches were collected, the acid became
dark olive, when 9 dark red, when 13 bright

orange, and when 18 pale. "It' had loft 31 grains, and when completely cool, was of fpecific gravity 1,502 nearly. The water in the apparatus was tinged of a light blue; from whence we may conclude that fome of the nitrous gas was abforbed by it with the nitrous acid: but it will be hereafter proved that the orange colored acid is the moft nitrated acid capable of combining undecompounded with water, and that the color it communicates to a large quantity of water, is light blue. If then we take 6,1 grains, the quantity of gas collected, from 31 the lofs, the remainder is 24,9, which reafoning from the fynthetical experiment, may be fuppofed to contain nearly 3 cubic inches of nitrous gas. Confequently, 94,25 grains of dark green acid, of fpecific gravity 1,475, are compofed of nearly 21 cubic inches, or 7,2 grains of nitrous gas, and 87,05 grains of pale nitrous acid, of 1,504.

VIII. Comparing the different fynthetical and analytical experiments, we may conclude with tolerable accuracy, that 92,75 grains of bright yellow, or ftandard acid of 1,5, are compofed of 2,75 grains of nitrous.gas, and 90 grains of nitric acid of 1,504; but 92,75 grains of ftandard acid contain 85,23 grains of nitrous acid, compofed of about 27,23 of oxygene, and 58, nitrous gas: now from 58, take 2,75, and the remainder 55,25, is the quantity of nitrous gas contained in 90 grains of nitric acid of 1,504; confequently, 100 grains of it are compofed of 8,45 water, and 91,55 true acid, containing 61,32 nitrous gas, and 30,23 oxygene; or 27,01 nitrogene, and 64,54 oxygene: and the nitrogene in nitric acid, is to the oxygene as 1 to 2,389.

IX. My ingenious friend, Mr. JAMES THOMSON, has communicated to me fome obfervations relating to the compofition of nitrous acid (that is, the orange-colored acid), from which he draws a conclufion which is, in my opinion, countenanced by all the facts we are in poffeffion of, namely, " that it ought " not to be confidered as a diftinct and lefs

" oxygenated ſtate of acid, but ſimply as nitric
" or pale acid, holding in ſolution, that is,
" looſely combined with, nitrous gas."*

It is impoſſible to call any ſubſtance a ſimple acid
that is incapable of entering undecompounded
into combination with the alkalies, &c; but it
will appear hereafter that the ſalts called in the

---

* In a letter to me, dated Oct. 28, 1799, after giving an
account of ſome experiments on the phlogiſtication of
nitric acid by heat and light, he ſays, "It was from an
" attentive examination of the manner in which the nitric
" acid was phlogiſticated in theſe experiments, that I was
" confirmed in the ſuſpicion I had long before entertained,
" of the real difference between the *nitrous* and *nitric* acids.
" It is not enough to ſhew that in the *nitrous* acid, (that is,
" the nitric holding nitrous gas in ſolution), the proportion
" of oxygene in the whole compound is leſs than that enter-
" ing into the compoſition of the nitric acid, and that it is
" therefore leſs oxygenated. By the ſame mode of reaſoning
" we might prove that water, by abſorbing carbonic acid
" gas, became leſs oxygenated, which is abſurd. Should
" any one attempt to prove (which will be neceſſary to ſub-
" ſtantiate the generally received doctrine) that the oxygene
" of the nitrous gas combines with the oxygene of the acid,
" and the nitrogene; in like manner, ſo that the reſulting acid,
" when nitrous gas is abſorbed by nitric acid, is a binary
" combination of oxygene and nitrogene, he would find it
" ſomewhat more difficult than he at firſt imagined; it ap-
" pears to me impoſſible. It is much more conſonant with

new nomenclature *nitrites*, cannot be directly formed. If, indeed, it could be proved, that the heat produced by the combination of nitrous acid with salifiable bafes, was the only cause of the partial decompofition of it, and that when this process was effected in such a way as to prevent increafe of temperature, no nitrous gas was liberated, the common

" experiment to, fuppofe that nitrous acid is nothing more
" than nitric acid holding nitrous gas in folution, which
" might in conformity to the principles of the French
" nomenclature, be called nitrate of nitrogene. The difficulty,
" and in fome cafes the impoffibility, of forming nitrites,
" arifes from the weak affinity which nitrous gas has for
" nitric acid, compared with that of other fubftances; and
" the decompofition of nitrous acid) that is, nitrate of
" nitrogene) by an alkaline or metallic fubftance, is perfectly
" analogous to the decompofition of any other nitrate, the
" nitrous gas being difplaced by the fuperior affinity of the
" alkali for the acid.

" Agreeable to this theory, the falts denominated
" *nitrites* are in fact triple falts, or ternary combinations of
" nitric acid, nitrous gas, and falifiable bafes."

This theory is perfectly new to me. Other Chemifts to whom I have mentioned it, have likewife confidered it as new. Yet in a fubfequent letter Mr. Thomfon mentions that he had been told of the belief of a fimilar opinion among the French Chemifts.

theory might have some foundation; but though dilute phlogisticated nitrous acid combines * with alkaline solutions without decompofition, yet no excefs of nitrous gas is found in the folid falt: it is either difengaged in proportion as the water is evaporated, or it abforbs oxygene from the atmofphere, and becomes nitric acid.

In proportion as the nitrous acids contain more nitrous gas, fo in proportion do they more readily give it out. From the blue green acid it is liberated flowly at the temperature of 50°, and from the green likewife on agitation. The orange-coloured and yellow acids do not require a heat above 200° to free them of their nitrous gas; and all the

---

* In fome experiments made on the nitrites of potafh, and of ammoniac, before I was well acquainted with the compofition of nitric acid, I found that a light olive-colored acid of 1,28, was capable of being faturated by weak folutions of potafh and ammoniac, without lofing any nitrous gas; but after the evaporation of the neutralifed folution, at very low temperatures, the falts in all their properties refembled *nitrates*.

colored acids, when exposed to the atmosphere absorb oxygene, and become by degrees pale.

If the nitrous vapour, i. e. such as is disengaged during the *denitration* of the colored acids, was capable of combining with the alkalies, it might be suppoſed a diſtinct acid, and called nitrous acid ; and the acids of different colors might be conſidered ſimply as compounds of this acid with nitric acid ; but it appears to be nothing more than a ſolution of nitric acid in nitrous gas, incapable of condenſation, undecompounded, and when decompounded and condenſed, conſtituting the dark green acid, which is immiſcible with water,† and uncombinable with the alkalies.‡

It ſeems therefore reaſonable, till we are in poſſeſſion of new lights on the ſubject, to conſider, with Mr. Thomſon, the deoxygenated or nitrous acids ſimply as ſolutious of nitrous gas

---

† As is evident from the curious appearance of the dark green ſpherules, repulſive both to water, and light green acid.

‡ That is, undecompounded.

in nitric acid, and as analogous to the folutions of nitrous gas in the fulphuric and marine acids, &c. and the falts called nitrites, ternary combinations, fimilar to the triple compounds compofed of fulphuric acid, metallic oxides, and nitrous gas.*

Suppofing the truth of thefe principles according to the logic of the French nomenclature, there is no acid to which the term nitrous acid *ought* to be applied; but as it has been ufed to fignify the acids holding in folution nitrous gas, it is perhaps better ftill to apply it to thofe fubftances, than to invent for them new names. A nomenclature, accurately expreffing their conftituent parts, would be too complex, and like all other nomenclatures founded upon theory, liable to perpetual alterations. Their compofition is known from their fpecific gravity and their colors; hence it is better to denote it by thofe phyfical properties: thus orange nitrous acid, of fpecific gravity 1,480, will fignify a folution of nitrous

* The exiftence of thefe bodies will be hereafter proved.

gas in nitric acid, in which the nitric acid is to the nitrous gas, nearly as 87 to 5, and to the water as 11 to 1.

X. The estimation of the composition of the yellow and orange colored nitrous acids given in the following table, may be considered as tolerably accurate, being deduced from the synthetical experiments in the sixth section, compared with the analytical ones. But as in the synthetical experiment, when the acid became green, it was impossible to ascertain the quantity of nitrous gas that passed through it unabsorbed, and as in the analysis the quantity of nitrous gas dissolved by the water at different periods of the experiment could not be ascertained, the accounts of the composition of the green acids must be considered only as very imperfect approximations to truth.

TABLE I.

Containing Approximations to the quantities of NITRIC ACID, NITROUS GAS, and WATER in NITROUS ACIDS, of different colors and specific gravities.

100 Parts		Specific gra.		NitricAcid	Water	Nitrous gas.
Sol. Nitric Acid		1,504		91,55	8,45	— —
YellowNitrous‡		1,502		90,5	8,3	1,2
Bright Yellow	of	1,500	contain	88,94	8,10	2,96
Dark Orange		1,480		86,84	7,6	5,56
Light Olive ‡		1,479		86,00	7,55	6,45
Dark Olive ‡		1,478		85,4	7,5	7,1
Bright Green ‡		1,476		84,8	7,44	7,76
Blue Green*		1,475		84,6	7,4	8,00

* The blue green acid is not homogeneal in its compofition, it is compofed of the blue green fpherules and the bright green acid. The blue green fpherules are of greater fpecific gravity than the dark green acid, probably becaufe they contain little or no water.

‡ The compofition of the acids thus marked, is given from calculations.

TABLE II.

Binary Proportions of OXYGENE and NITROGENE in NITRIC and NITROUS ACIDS.*

100 Parts.		Oxy-gene	Nitro-gene	Proportions. Nitrogene. Unity.	Nitro-gene	Oxy-gene
Nitric Acid	contain	70,50	29,50		1	2,389
Bright yellow Nitrous		70,10	29,90		1	2,344
Orange coloured		69,63	30,37		1	2,292
Dark Green		69,08	30,92		1	2,230

XI. I have before mentioned that dilute nitric acids are incapable of diffolving fo much nitrous gas in proportion to their quantities of true acid, as concentrated ones. During their abforption of it, they go through fimilar changes of color; 330 grains of nitric acid, of fpecific gravity 1,36, after 50 cubic inches of gas had been paffed through it, became blue green, and

* Nitrous gas contains 44,05 Nitrogene, and 55,95 Oxygene, as has been faid before.

of fpecific gravity 1,351. It had gained in weight but 3 grains; and when the nitrous gas was driven from it by heat into a water apparatus, but 7 cubic inches were collected.*

From the diminution of fpecific gravity of nitric acid by combination with nitrous gas, and from the fmaller attraction of nitric acid for nitrous gas, in proportion as it is diluted, it is probable that the nitrated acids, in their combinations with water, do not contract fo much as † nitric acids of the fame fpecific gravities. The affinities refulting from the fmall attraction of nitrous gas for water, and its greater attraction for nitric acid, muft be fuch as to leffen the affinity of nitric acid and water for each other.

Hence it would require an infinite number of experiments to afcertain the real quantities of acid, nitrous gas, and water, contained in the

---

* A great portion of it, of courfe, diffolved in the water with the nitrous acid carried over.

† Their changes of volume, correfponding to changes of temperature, moft probably, are likewife different.

different diluted nitrous acids; and after these quantities were determined, they would probably have no important connection with the chemical arrangement. As yet, our instruments of experiment are not sufficiently exact to afford us the means of ascertaining the ratio in which the attraction of nitric acid* for water diminishes in its progress towards saturation.

The estimations in the following table, of the real quantities of nitric acid in solutions of different specific gravities, were deduced from experiments made in the manner described in section VI, except that the phial employed was longer, narrower, and graduated to half grains. The temperature, at the time of combination, was from 40° to 46°.

---

* Probably in the ratio of the square of the quantity of water united to it,

TABLE III.

Of the Quantities of True NITRIC ACID in solutions of different SPECIFIC GRAVITIES.

100 Parts Nitric Acid of fpecfic gravity		True Acid*	Water
1,5040		91,55	8,45
1,4475		80,39	19,61
1,4285	contain	71,65	28,35
1,3906		62,96	37,04
1,3551		56,88	43,12
1,3186		52,03	47,97
1,3042		49,04	50,96
1,2831		46,03	53,97
1,2090		45,27	54,73

* The quantities of Oxygene and Nitrogene in any folution, may be thus found —— Let A = the true acid,
X the oxygene, and Y the nitrogene,

$$\text{Then} \quad X = \frac{238\,A}{239} \quad \text{and} \quad Y = \frac{A}{239}$$

XII. The blue green fpherules mentioned in fe&ion V. produced by the condenfation of nitrous vapor, and by the combination of nitric acid with nitrous gas, may be confidered as faturated folutions of nitrous gas in nitric acid. The combinations of nitric acid and nitrous gas containing a larger proportion of nitrous gas, are incapable of exifting in the fluid ftate at common temperatures; and, as appears from the firft experiment, an increafe of volume take place during their formation. They confequently ought to be looked upon as folutions of nitric acid in nitrous gas, identical with the nitrous vapor of Prieftley.

From the refearches of this great difcoverer, we learn that nitrous vapor is decompofable, both by water and mercury. Hence it is almoft impoffible accurately to afcertain its compofition. In one of his experiments,‡ when more than 130 grains of ftrong nitrous acid were expofed

‡ Experiments and Obfervations; laft edition, vol. 1, page 384.

for, two days to nearly 247 cubic inches of nitrous gas, over water: about half of the acid was diffolved, and depofited with the gas in the water.§

XIII. In comparing the refults of my fundamental experiment on the compofition of nitrous acid, with thofe of Cavendifh, the great coincidence between them gave me very high fatisfaction, as affording additional proofs of accuracy. If the acid formed in the laft experiment of this illuftrious philofopher be fuppofed analogous to the light green acid formed in my firft experiment, our eftimations will be almoft identical.

Lavoifier's account of the compofition of the nitric and nitrous acids, has been generally adopted. According to his eftimation, thefe fubftances contain a much larger quantity of oxygene than I have affigned to them.

§ Nitrous gas, holding in folution nitrous acid, is more readily abforbed by water than when in its pure form, from being prefented to it in a more condenfed ftate in the green acid, formed by the contact of water and nitrous vapor.

·'The fundamental experiments of this great phi-
lófopher were made at an early period of pneumatic chemiftry,* on the decómpofition of niire
by charcoal; and he confidered the nitrogene
evolved, and the oxygene of the carbonic acid
produced in this procefs, as the component
parts of the nitric acid contained in the nitre.

I have before mentioned the liberation of
nitrous acid, in the decompofition of nitre by
combuftible bodies; and I had reafons for fuf-
pecting that this circumftance was not the only
fource of inaccuracy.

That my fufpicions were well founded, will
appear from the following experiments :

EXPERIMENT *a.* I introduced into a
ftrong glafs tube, 3 inches long, and nearly ,8
wide, a mixture of 10 grains of pulverifed,
well burnt charcoal, and 60 grains of nitre. It
was fired by means of touch-paper, and the
tube inftantly plunged under a jar filled with

* Mem. des Savans Etrangers, v. xi. 226. Vide Kirwan
fur le phlogiftique pag. 110.

dry mercury. A quantity of gas, clouded with denfe white vapor was collected. When this vapor was precipitated, fo that the furface of the mercury could be feen, it appeared white, as if acted on by nitrous acid. On introducing a little oxygene into the jar, copious red fumes appeared.

EXP. *b.* A fimilar mixture was fired* under the jar, the top of the mercury being covered with a fmall quantity of red cabbage juice, rendered green by an alkali. This juice, examined when the vapor was precipitated, was become red, and on introducing to it a little carbonate of potafh, a flight effervefcence took place.

EXP. *c.* Five grains of charcoal, and 20 of nitre, were now fired in the fame manner as before, the mercury being covered with a ftratum of water. After the precipitation of the vapor

---

* In this experiment, as well as in the laft, fome of the mixture was thrown into the jar undecompounded.

on the introduction of oxygene, no red fumes were perceived.

EXP. *d.* 30 grains of nitre, 5 of charcoal, and five of filicious earth,* were now mingled and fired. The gas received under mercury was compofed of 18 carbonic acid, and nearly 12 nitrogene.† A little muriatic acid was poured on the refiduum in the tube; a flight effervefcence took place.

EXP. *e.* The top of the mercury in the jar was now covered with a little diluted muriatic acid, and a fmall glafs tube filled with a mixture of 3 grains of charcoal, and 20 nitre. After the deflagration, the tube itfelf with the refiduum it contained, were thrown into the jar. The carbonic acid was quickly detached from them by the muriatic acid, and the whole quan-

---

* To detach the potafh from the carbonic acid.

† This nitrogene contained a little nitrous gas, as it gave red fumes when expofed to the air. The free nitrous acid was decompofed by the mercury, as it was not covered with water.

tity of gas generated in the procefs, obtained; it meafured 15 cubic inches.

4 cubic inches of it expofed to folution of potafh, diminifhed to $1\frac{4}{10}$; 7 of the remainder, with 8 of oxygene, gave only 12.

EXP. *f.* 60 grains of nitre, and 9 of charcoal were fired, the top of the mercury in the jar being covered with water. After the deflagration, the tube that had contained them was introduced, and the carbonic acid contained by the carbonate of potafh, difengaged by muriatic acid. 30 meafures of the gafes evolved were expofed to caustic potafh; 20 exactly were abforbed, the 10 remaining, with 10 of oxygene, diminifhed to 17.

EXP. *g.* A mixture of nitre and charcoal were deflagrated over a little water in the mercurial jar: after the precipitation of the vapor, the water was abforbed by filtrating paper. This filtrating paper, heated in a folution of potafh, gave a faint fmell of ammoniac.

EXP. *h.* Water impregnated with the vapor produced in the deflagration, was heated with quicklime, and prefented feparately to three perfons accuftomed to chemical odors. Two of them inftantly recognifed the ammoniacal fmell, the other could not afcertain it. Paper reddened with cabbage juice was quickly turned green by the vapor.

Thefe experiments are fufficient to fhew that the decompofition of nitre by charcoal is a very complex procefs, and that the intenfe degree of heat produced may effect changes in the fubftances employed, which we are unable to eftimate.

The products, inftead of being fimply carbonic acid, and nitrogene, are carbonic acid, nitrogene, nitrous acid, probably ammonia, and fometimes nitrous gas. The nitrous acid is difengaged from the bafe by the intenfe heat. Concerning the formation of the ammonia, it is ufelefs to reafon till we have obtained unequivocal teftimonies of its exiftence; it may be produced either by the decompofition of the water contained in the nitre, by the combination of its oxygene with the charcoal, and of its nafcent hydrogene with the nitrogene of

the nitric acid; or from fome unknown decompofition of the potafh. ..

As neither Lavoifier, nor Berthollet found
nitrous gas produced in the decompofition of
nitre by charcoal, when a water apparatus was
employed; and as it was not uniformly evolved
in my experiments, the moft probable fuppofition is, that it arifes from the decompofition
of a portion of the free nitrous acid intenfely
heated, by the mercury.

In none of my experiments was the whole of
the nitre and charcoal decompofed, fome of it
was uniformly thrown with the gafes into the
mercurial apparatus. The nitrogene evolved,
as far as I could afcertain by the common tefts,
was mingled with no inflammable gas.

If we confider experiment *f* as accurate, with
regard to the relative quantities of carbonic acid
and nitrogene produced, they are to each other
nearly as 20 to 8; that is, allowing 2, for the
nitrous gas, and confequently, reafoning in the
fame manner as Lavoifier, concerning the compofition of nitric acid, it fhould be compofed

D

of 1 nitrogene to 3,38 oxygene. But though the quantity of oxygene in this eftimation is far fhort of that given in his, yet ftill it is too much. From whatever fource the errors arife, whether from the evolution of phlogifticated nitrous acid, or the decompofition of water, or the production of nitrous gas, they all tend to increafe the proportion of the carbonic acid to the nitrogene.

I am unacquainted with any experiment from which accurate opinions concerning the different relative proportions of oxygene and nitrogene in the nitric and nitrous acids could be deduced. Lavoifier's calculation is founded on his fundamental experiment, and on the combination of nitrous gas and oxygene.

Dr. Prieftley's experiment mentioned in fection 12, on the abforption of nitrous gas by nitrous acid, from which Kirwan* deduces the compofition of the differently colored nitrous acids, was made over water, by which, as is

---

* Effay on phlogifton.

evident from a minute examination of the facts†, the greater portion of the nitrous gas employed was abforbed.

XIV. The opinions heretofore adopted refpecting the quantities of real or true acid in folutions of nitrous acid of different fpecific gravities, have been founded on experiments made on the nitro-neutral falts, the moft accu-

---

† Dr. Prieftley fays, " Having filled a phial containing " exactly the quantity of four pennyweights of water, with " ftrong, pale, yellow fpirit of nitre, with its mouth quite " clofe to the top of a large receiver ftanding in water, I " carefully drew out almoft all the common air, and then " filled it with nitrous air; and as this was abforbed, I kept " putting in more and more, till in lefs than two days it " had completely abforbed 130 ounce meafures. Prefently " after this process began, the furface of the acid affumed " a deep orange color, and when 20 or 30 ounce meafures " of air were abforbed, it became green at the top: this " green defcended lower and lower, till it reached the " bottom of the phial. Towards the end of the procefs, " the evaporation was perceived to be very great, and when " I took it out, the quantity was found to have diminifhed " to one half. Alfo it had become, by means of this pro- " cefs, and the evaporation together, exceeding weak, and " was rather blue than green."

Experiments and Obfervations, vol. 1, p. 384. Laft edition.

rate, of which are thofe of Kirwan, Bergman, and Wenzel. The great difference in the refults of thefe celebrated men, proves the difficulty of the inveftigation, and the exiftence of fources of error.* Kirwan deduces the compofition of the folutions of nitrous acid in water, from an experiment on the formation of nitrated foda. In this experiment, 36,05 grains of foda were faturated by 145 grains of nitrous acid, of fpecific gravity 1,2754. By a teft experiment, he found the quantity of falt formed to be 85,142 grains.† Hence he concludes that 100 parts of nitrous acid, of fpecific gravity 1,5543, contain 73,54 of the ftrongeft, or moft concentrated acid.

Suppofing his eftimation perfectly true, 100 parts of the aëriform acid of 55° would be com-. pofed of 74,54 of his real acid, and 25,46 water. In examining, however, one of his later

---

* See Mr. Keir's excellent obfervations on this fubject. Chem. Dict. Art. Acid.

† Irifh Tranfactions, vol. 4, p. 34.

experiments,* we shall find reasons for conclu-
ding, that the acid in nitrated soda cannot contain much less water than the aëriform acid. A
solution of carbonated soda, containing 125
grains of real alkali, was saturated by 306,2
grains of nitrous acid, of specific gravity 1,416.
The evaporation was carried on in a temperature
not exceeding 120°, and the residuum exposed
to a heat of 400° for six hours, at the end of
which time it weighed 308 grains. Now according to my estimation, 306 grains of nitric
acid, of 1,416, should contain 215 true acid;
and we can hardly suppose, but that during the
evaporation and consequent long exposure to
heat, some of the nitrated soda was lost with
the water.

Bergman estimates the quantity of water in
this salt at 25, and the acid at 43 per cent; but
his real acid was not so concentrated as Kirwan's, consequently the nitric acid in nitrated soda should contain more water than my
true acid.

* Addit. Obs. pag. 74.

Wenzel, from an experiment on the compofition of nitrated foda, concludes that it contains 37,48 of alkali, and 62,52 of nitrous acid;
and 1000 of this acid, from Kirwan's calculation, contain 812,6 of his real acid; confequently, 100 parts of my aëriform acid fhould
contain 93,28 of Wenzel's acid, and 6,72 of
water.

I faturated with potafh 54 grains of folution
of nitric acid, of fpecific gravity 1,301. Evaporated at about 212°, it produced 66 grains of
nitre. This nitre expofed to a higher temperature, and kept in fufion for fome time, was
reduced to 60 grains.

Now from the table, 54 of 1,301, fhould
contain 26,5 of true acid. But according to
Kirwan's eftimation, 100 parts of dry nitre
contain 44* of his real acid, with 4 water;
confequently 60 fhould contain 26,4.

Again, 90 grains of acid, of fpecific gravity
1,504, faturated with potafh, and treated in

* Additional Obfervations, page 79.

the fame manner, gave 173 grains of dry nitre. Confequently, 100 parts of it fhould contain 47,3 grains of true acid.

Now Lavoifier † allows about 51 of dry acid to 100 grains of nitre, and Wenzel 52.

From Berthollet's‡ experiments, 100 grains of nitre, in their decompofition by heat, give out nearly 49 grains of gas.§

Hence it appears that the aëriform acid, that is, the true acid of my table, contains rather lefs water than the acid fuppofed to exift in nitre.

---

† Elements, pag. 103, Kerr's Tranflation.

‡ Mem. Acad. 1787.

§ As well as oxygene and nitrogene, Mr. Watt's experiments prove that much phlogifticated nitrous acid is produced.

DIVISION II.

EXPERIMENTS and OBSERVATIONS on the com-
position of AMMONIAC and on its combinations with
WATER and NITRIC ACID.

I. Analysis of AMMONIAC or VOLATILE ALKALI.

THE formation and decompofition of volatile
alkali in many proceffes, was obferved by Prieftley, Scheele, Bergman, Kirwan, and Higgins ;
but to Berthollet we owe the difcovery of its
conftituent parts, and their proportions to each
other. Thefe proportions this excellent philofopher deduced from an experiment on the
decompofition of aëriform ammoniac by the
electric fpark :* a procefs in which no apparent
fource of error exifts.

* Journal de Phyfique. 1786. Tom. 2, pag. 176.

Since, however, his eftimations have been made, the proportions of oxygene and hydrogene in water have been more accurately determined. This circumftance, as well as the conviction of the impoffibility of too minutely fcrutinizing facts, fundamental to a great mafs of reafoning, induced me to make the following experiments.

A porcelain tube was provided, open at both ends, and well glazed infide and outfide, its diameter being about ,5 inches. To one end of this, a glafs tube was affixed, curved for the purpofe of communicating with the water apparatus. With the other end a glafs retort was accurately connected, containing a mixture of perfectly cauftic flacked lime, and muriate of ammoniac.

The water in the apparatus for receiving the gafes had been previoufly boiled, to expel the air it might contain, and during the experiment was yet warm.

When the tube had been reddened in a furnace adapted to the purpofe, the flame of a ſpirit lamp was applied to the bottom of the retort. A great quantity of gas was collected in the water apparatus; of this the firſt portions were rejected, and the laſt transferred to the mercurial trough.

A ſmall quantity examined, did not at all diminiſh with nitrous gas, and burnt with a lambent white flame, in contact with common air.

$2\frac{3}{4}$ of this gas, equal to 110 grain meaſures, were fired with 2, equal to 80, of oxygene, in a detonating tube, by the electric ſpark. They were reduced to $2\frac{1}{4}$, or 90. On introducing to the remainder a ſolution of ſtrontian, it became ſlightly clouded on the top, and an abſorption of ſome grain meaſures took place.

It was evident, then, that in this experiment, charcoal * had been ſomehow preſent in the

---

* Though the tube had never been uſed, and was apparently clean and dry on the inſide, it muſt have contained ſomething in the form of duſt, capable of furniſhing either hydro-carbonate, or charcoal.

tube; which being diffolved by the nafcent hydrogene, had rendered it flightly carbonated, and in confequence made the refults inconclufive.

A tube of thick green glafs carefully made clean, was now employed, inclofed in the porcelain tube. Every other precaution was taken to prevent the exiftence of fources of error, and the experiment conducted as before.

140 grain meafures of the gas produced, fired with 120 of oxygene, left, in two experiments, nearly 110. Solution of ftrontian placed in contact with the refiduum, did not become clouded, and no abforption was perceived.

Now 150 meafures of gas were deftroyed, and if we take Lavoifier's and Meufnier's eftimation of the compofition of water, and fuppofe the weight of oxygene to be 35 grains, and that of hydrogene 2,6 the hundred cubic inches; the oxygene employed will be to the hydrogene as 243 to 576. Put x for the oxygene, and y for the hydrogene.

$$x : y :: 243 : 576,$$

$$x = \frac{243\,y}{576}$$

$$819\,y = 86400$$

$$y = 105 \qquad x = 45$$

And $140 - 105 = 35$

Confequently, the nitrogene in ammoniac is to the hydrogene as 35 : 105 in volume : and 13,3 grains of ammoniac are compofed of 10,6 nitrogene, (fuppofing that 100 cubic inches weigh 30,45 grains) and 2,7 hydrogene.

According to Berthollet, the weight of the nitrogene in ammoniac is to that of the hydrogene as 121 to 29.* The difference between this eftimation and mine is fo fmall as to be almoft unworthy of notice, and arifes moft probably from the flight difference between the accounts of Lavoifier and Monge, of the compofition of water, and the different weights affigned to the gafes employed.

* Journal de Phyfique, 1786, t. 2, 177.

We may then conclude, that 100 grains of ammoniac are compofed of, about 80 nitrogene, and 20 hydrogene.,

The decompofition of ammoniac by heat, as well as by the electric fpark, was firft difcovered by Prieftley. In an experiment† when aëriform ammoniac was fent through a heated tube from a cauftic folution of ammoniac in water, this great difcoverer obferved that an inflammable gas was produced, though in no great quantity, and that a fluid blackened by matter, probably carbonaceous, likewife came over.

In my experiments the whole of the ammoniac appeared to be decompofed; the quantity of gas generated was immenfe, and not clouded, as is ufually the cafe with gafes generated at high temperatures. It is poffible, that the larger quantity of water carried over in his experiment, by its ftrong attraction for ammoniac in the aëriform ftate, might have, in fome meafure, retarded the decompofition. It is how-

* Phil. Tranf. vol. 79, page 294.

ever; more probable to fuppofe, that a fiffure exifted in the earthen tube he employed, through which a certain quantity of gas efcaped, and coaly matter entered.

Prieftley found that the metallic oxides when ftrongly heated, decompofed ammoniac, the metal being revivified and water and nitrogene produced.* The eftimations of the compofition of ammoniac that may be deduced from his experiments on the oxide of lead, differ very little from thofe already detailed.

II. *Specific gravity of Ammoniac.*

From the great folubility of ammoniac in water, it is difficult to afcertain its fpecific gravity in the fame manner as that of a gas combinable to no great extent with that fluid. It is impoffible to prevent the exiftence of a

fmall quantity of folution of ammoniac in the
mercurial airholder,† or apparatus containing
thé gas; and during the diminution of the
preffure of the atmofphere on this folution,‡ a
certain quantity of gas is liberated from it, and
hence a fource of error.

To afcertain, then, the weight of ammoniac,
I employed an apparatus fimilar to that ufed
for the abforption of nitrous gas by nitric
acid.

50 cubic inches of gas were collected in the
mercurial airholder, from the decompofition of
muriate of ammoniac by lime; thermometer
being 58°, and barometer 29,6.

100 grains of diluted fulphuric acid were
introduced into the fmall graduated cylinder,
which after being carefully weighed, was made

---

† Ammoniac generated at a temperature above that of
the atmofphere, always depofits ammoniacal folution during
its reduction to the common temperature.

‡ By the introduction of aëriform ammoniac into the
exhaufted globe.

to communicate with the airholder, the curved tube containing a small quantity of water. The gas was slowly passed into the fluid, and the globules wholly absorbed, before they reached the top; much increase of temperature being consequent. When the absorption was compleat, the phial was increased in weight exactly 9 grains.

This experiment was repeated three times. The difference of weight, which was probably connected with alterations of temperature and pressure, never amounted to more than one fixth of a grain.

We may then conclude, that at temperature 58°, and atmospheric pressure 29,6, 100 cubic inches of ammoniac weigh 18 grains.

According to Kirwan, 100 cubic inches of alkaline air * weigh 18,16 grains; barometer 30,0, thermometer 61. The difference between these estimations, the corrections for temperature and pressure being made, is trifling.

* Additional Obfervations, page 107.

III. *Of the quantities of true Ammoniac in Aqueous Ammoniacal Solutions, of different specific gravities.*

To ascertain the quantities of ammoniac, such as exists in the aëriform state, saturated with moisture, in solutions of different specific gravities, I employed the apparatus for absorption so often mentioned. Thermometer being 52°, the mercurial airholder was filled with ammoniacal gas, and the graduated phial, containing 50 grains of pure water, connected with it. During the absorption of the gas, the phial became warm. When about 30 cubic inches had been passed through, it was suffered to cool, and weighed : it had gained 5,25 grains, and the fluid filled a space equal to that occupied by 57* grains of water.

---

* It is necessary in these experiments, that the greatest care be observed in the introduction and extraction of the capillary tube. If it is introduced dry, there will be a source of error from the moisture adhering to it when taken out. I therefore always wetted it before its introduction, and took care that no more fluid adhered to it after the experiment, than before.

E

Confequently, 100 grains of folution of ammoniac in water of fpecific gravity ,9684 contain 9,502 grains of ammoniac.

The apparatus being adjufted as before, 50 grains of pure water were now perfectly faturated with ammoniac. They gained in weight 17 grains, and when perfectly cool, filled a fpace equal to 74 of water Confequently 100 grains of aqueous ammonial folution of fpecific gravity ,9054 contain 25,37 grains of ammoniac.

The two folutions were mingled together; but no alteration of temperature took place. Confequently the refulting fpecific gravity might have been found by calculation.

On mingling a large quantity of cauftic folution of ammoniac with $\frac{1}{4}$ of its weight of water, of exactly the fame temperature, no alteration of it was perceptible by a fenfible thermometer.— Hence the two experiments* being affumed as

* Previous to thofe experiments, I had made a number of others on the combination of ammoniac with water.— My defign was, to afcertain the diminution of fpecific

data, the intermediate eftimations in the following table, were found by calculation.

gravity for every three grains of ammoniac abforbed; but
this I found impoffible. The capillary tube, when taken
out of the phial, always carried with it a minute portion
of the folution, which partially evaporated before it could
be again introduced; and thus the fources of error increafed
in proportion to the number of examinations.

TABLE IV.

Of approximations to the quantities of AMMONIAC, such as exists in the aëriform state, saturated with water at 52°, in AQUEOUS AMMONIACAL SOLUTIONS of different specific gravities.

100 Specific gra.		Ammoniac	Water.
9054		25,37	74,63
9166		22,07	77,93
9255		19,54	80,46
9326		17,52	82,48
9385		5,88	84,12
9435		14,53	85,47
9476	contain	13,46	86,54
9513		12,40	87,60
9545		11,56	88,44
9573		10,82	89,18
9597		10,17	89,83
9619		9,60	90,40
9684		9,50	90,5
9639		9,09	90,91
9713		7,17	92,83

* As yet no mode has been difcovered for obtaining gafes in a ftate of abfolute drynefs; confequently we are ignorant of the different quantities of water they hold in folution at different temperatures. As far as we are acquainted with the combinations of ammoniac, there is no ftate in which it exifts fo free from moifture, as when aëriform, at low temperatures.

That no confiderable fource of error exifted in the two experiments, is evident from the trifling difference between the eftimations of the quantities of real ammoniac, in the folution of ,9684, as found in the firft experiment, and as given by calculation from the laft.

The quantity of ammoniac in a folution of fpecific gravity not in the table, may be thus determined—Find the difference between the two fpecific gravities neareft to it in the table; d, and the difference between their quantities of alkali, b; likewife the difference between the given fpecific gravity and that neareft to it, c.

then $d : b :: c : x$ and $x = \dfrac{b \, c}{d}$

Which, added to the quantity of the lower fpecific gravity, is the alkali fought.

The differences in fpecific gravity of the folutions of ammoniac at temperatures between 40° and 65° * are fo trifling as to be hardly

* The expanfion from increafe of temperature is probably great in proportion to the quantity of ammoniac in the folution.

ascertainable, by our imperfect instruments, and consequently are unworthy of notice.

It is possible at very low temperatures to obtain ammoniacal solutions of less specific gravity than ,9, but they are incapable of being kept for any length of time under the common pressure of the atmosphere.

IV. *Combinations of Ammoniac with Nitric Acid. Composition of Nitrate of Ammoniac, &c.*

200 grains of ammoniacal solution, of specific gravity ,9056, were saturated by 385,5 grains of nitric acid, of specific gravity 1,306. The combination was effected in a long phial, the nitrous acid added very slowly, and the phial closed after every addition, to prevent any evaporation in consequence of the great increase of temperature.† The specific gravity of the solution, when reduced to the common temperature, was 1,15. Evaporated at a heat of

---

† From the combination.

212°,‡ it gave 254 grains of falt of fibrous cryftalization. This falt was diffolved in 331 grains of water; the fpecific gravity of the folution was 1,148 nearly.

Hence it was evident that fome of the falt had been loft during the evaporation.

To find the quantity loft, fibrous nitrate of ammoniac was diffolved in fmall quantities in the folution, the fpecific gravity of which, was examined after every addition of 3 grains. When 16 grains had been added to it, it became of 1,15.

Confequently, the folution compofed of 200 grains of ammoniacal, and of 385,5 of nitric acid folution, contained 262 grains of falt of fibrous cryftalization, and of this falt 8 grains were loft during the evaporation.

But the alkali in 200 grains of ammoniacal folution of ,9056 = 50,5 grains. And the true nitric acid in 385,5 grains of folution of 1,306 = 190 grains.

‡ I had before proved that at this temperature the falt neither decompofed nor fublimed.

water.

And 262 grains of fibrous cryftalized nitrate of ammoniac, contain 190 grains true acid, 50,5 ammoniac, and 21,5 water. And 100 parts contain 72,5 acid, 19,3 ammoniac, and 8,2 water.

In proportion as the temperature employed for the evaporation of nitro-ammoniacal folutions, is above or below 212°, fo in proportion does the falt produced contain more or lefs water than the fibrous nitrate. But whatever may have been the temperature of evaporation, the acid and alkali appear always to be in the fame proportions to each other.

Of the falts containing different quantities of water, two varieties muft be particularly noticed. The prifmatic nitrate of ammoniac, produced at the common temperatures of the atmofphere, and containing its full quantity of water of cryftalifation; and the compact nitrate of ammoniac, either amorphous, or compofed of delicately needled cryftals, formed at 300°, and containing but little more water, than exists in nitric acid and ammoniac.

2. To discover the composition of the prismatic nitrate of ammoniac, 200 grains of fibrous salt were dissolved in the smallest possible quantity of water, and evaporated in a temperature not exceeding 70°. The greater part of the salt was composed of perfectly formed tetrahædral prisms, terminated by tetrahædral pyramids. It had gained in weight about 8,5 grains.

Consequently 100 grains of prismatic nitrate of ammoniac may be supposed to contain 69,5 acid, 18,4 ammoniac, and 12,1 water.

To ascertain the composition of the compact nitrate of ammoniac, I exposed in a deep porcelain cup, 400 grains of the fibrous salt, in a temperature below 300°. It quickly became fluid, and slowly gave out its water without any ebullition, or liberation of gas. When it was become perfectly dry, it had lost 33 grains. I suspected, that in this experiment some of the salt had been carried off with the water; to determine this, I introduced into a small glass

retort; 460 grains of fibrous falt ; it was kept
at a heat below 320°, in communication with
a mercurial apparatus, in a regulated air-furnace, till it was perfectly dry : it had loft
23 grains. No gas, except the common air of
the retort came over, and the fluid collected
had but a faint tafte of nitrate of ammoniac.

Though in this experiment I had removed
all the fluid retained in the neck of the
retort, ftill a few drops remained in the head,
and on the fides, which I could not obtain. It
was of importance to me to be accurately acquainted with the compofition of the compact
falt, and for that reafon I compared thefe analytical experiments with a fynthetical one.

I faturated 200 grains of folution of ammoniac, of ,9056 with acid, afcertained the fpecific gravity of the folution, evaporated it at
212°, and fufed and dried it at about 300°—
260°. It gave 246 grains of falt, and a folution made of the fame fpecific gravity as that
evaporated, indicated a lofs of 9 grains. Confequently, 255 grains of this falt contain 50,5

grains alkali, 190 grains acid, and 145 grains water.

We may then conclude, that 100 parts of compact nitrate of ammoniac contain 74,5 acid, 19,8 alkali, and 5,7 water.

V. *Decomposition of Carbonate of Ammoniac by Nitric Acid.*

In my first experiments on the production of nitrate of ammoniac, I endeavoured to ascertain its composition by decompounding carbonate of ammoniac by nitric acid; and in making for this purpose, the analysis of carbonate of ammoniac, I discovered that there existed many varieties of this salt, containing very different proportions of carbonic acid, alkali, and water; the carbonic acid and water being superabundant in it, in proportion as the temperature of its formation was low, and the alkali in proportion as it was high: and not only that a different salt was formed at every different temperature, but likewise that the difference in

them was fo great; that the carbonate of ammoniac formed at 300° contained more than 50
per cent alkali, whilft that produced at 60°, contained only 20.*

I found 210 grains of carbonate of ammoniac, which from comparifon with other falts
previoufly analifed, I fufpected to contain about
20 or 21 per cent alkali, faturated by 200
grains of nitric acid of 1,504. But though
the carbonate was diffolved in much water,
ftill, from the fmell of the carbonic acid generated, I fufpect that a fmall portion of the
nitric acid was diffolved, and carried off by
it. The folution, evaporated at about 200°,
and afterwards expofed to a temperature below
300°, gave 232 grains of compact falt. But
reafoning from the quantity of acid in 200
grains of nitric acid of 1,504, it ought to have
given 245. Confequently 13 were loft by

* A particular account of the experiments from which
thefe facts were deduced, was printed in September, and
will appear in the firft volume of the *Refearches*.

evaporation; and this loſs agrees with that in the other experiments.

V. *Decompoſition of Sulphate of Ammoniac by Nitre.*

As a cheap mode of obtaining nitrate of ammoniac, Dr. BEDDOES propoſed to decompoſe nitre by ſulphate of ammoniac, which is a well known article of commerce. From ſyntheſis of ſulphate of ammoniac, compared with analyſis made in Auguſt 1799,* I concluded that 100 grains of priſmatic ſalt were compoſed of about 18 grains ammoniac, 44 acid, and 38 water; and ſuppoſing 100 grains of nitre to contain 50 acid, 100 grains of ſulphate of ammoniac will require for their decompoſition 134 grains of nitre, and form 90,9 grains of compact nitrate of ammoniac.

---

* And which will be publiſhed, with an account of its perfect decompoſition at a high temperature, in the *Reſearches.*

To ascertain if the sulphate of potash and nitrate of ammoniac could be easily separated, I added to a heated saturated solution of sulphate of ammoniac, pulverised nitre, till the decomposition was complete. After this decomposition, the solution contained a slight excess of sulphuric acid, which was combined with lime, and the whole set to evaporate at a temperature below 250°. As soon as the sulphate of potash began to cryftalise, the solution was suffered to cool, and then poured off from the cryftalised salt, which appeared to contain no nitrate of ammoniac. After a second evaporation and cryftalisation, almost the whole of the sulphate appeared to be deposited, and the solution of nitrate of ammoniac was obtained nearly pure: it was evaporated at 212°, and gave fibrous cryftals.

VI. *Non-exiftence of Ammoniacal Nitrites.*

I attempted in different modes to combine *nitrous* acids with ammoniac, so as to form the falts which have been fuppofed to exift, and

called *nitrites* of ammoniac ; but without fuccefs.

I firſt decompoſed a folution of carbonate of
ammoniac by dilute olive colored acid ; but in
this procefs, though no heat was generated,
yet all the nitrous gas appeared to be liberated
with the carbonic acid.* I then combined a
fmall quantity of nitrous gas, with a folution
of nitrate of ammoniac. But after evaporating
this folution at 70°—80°, I could not detect
the exiſtence of nitrous gas in the folid falt ;
it was given out during the evaporation and
cryſtalifation, and formed into nitrous acid by
the oxygene of the atmoſphere. I likewife
heated nitrate of ammoniac to different degrees,
and partially decompoſed it, to afcertain if in
any cafe the acid was phlogiſticated by heat :
but in no experiment could I detect the exiſtence

---

* When nitrous gas exiſts in neutro-faline folutions,
they are always colored more or lefs intenfely, from yellow
to olive, in proportion to the quantity combined with
them.

of *nitrous* acid in the heated falt, when it had been previoufly perfectly neutralifed.

When nitrate of ammoniac, indeed, with excefs of nitric acid, is expofed, to heat, the fuperabundant nitric acid becomes phlogifticated, and is then liberated from the falt, which remains neutral.*

We may therefore conclude that nitrous gas has little or no affinity for folid nitrate of ammoniac, and that no fubftance exifts to which the name *nitrite* of ammoniac can with propriety be applied.

VII. *Of the fources of error in Analyfis.*

To compare my fynthefis of nitrate of ammoniac with analyfis, I endeavoured to feparate the ammoniac and nitric acid from each other, without decompofition. But in going through the analytical procefs, I foon difcovered that

---

* Hence a nitrate of ammoniac with excefs of acid, when expofed to heat, firft becomes yellow, and then white.

it was impoffible to make it accurate, without many collateral laborious experiments on the quantities of ammoniac foluble in water at different temperatures.

At a temperature above 212°, I decompofed; by cauftic flacked lime, 50 grains of compact nitrate of ammoniac in a retort communicating with the mercurial airholder, the moifture in which had been previoufly faturated with ammoniac. 22 cubic inches of gas were collected at 38°, and from the lofs of weight of the retort, it appeared that 13 grains of folution of ammoniac in water, had been depofited by the gas.

Now evidently, this folution muft have contained much more alkali in proportion to its water than that of 55°, otherwife the quantity of ammoniac in 50 grains of falt would hardly equal 8 grains.*

---

* The accounts given by different chemifts of the compofition of nitrate of ammoniac, are extremely difcordant; they have been chiefly deduced from decompofitions of carbonate of ammoniac (the varieties of which have been

VIII. *Of the loſs of Solutions of Nitrate of Ammoniac during evaporation.*

. The moſt concentrated ſolution of nitrate of ammoniac capable of exiſting at 60°, is of ſpecific .gravity 1,304, and contains 33 water, and 67 fibrous ſalt, per cent. When this ſolution is evaporated at temperatures between 60º and 100, the ſalt is increaſed in weight by the addition of water of cryſtaliſation, and no portion of it is loſt..

During the evaporation of ſolutions of ſpecific gravity 1,146 and 1,15, at temperatures below 120°, I have never detected any loſs of ſalt. When the temperature of evaporation is 212º, the loſs is generally from 3 to 4 grains per cent; and when from 230° to the ſtandard of their ebullition, from 4 to 6 grains.

heretofore unknown) by nitrous acids of unknown degrees of nitration. Hence they are particularly erroneous with regard to the alkaline part. Wenzel ſuppoſes it to be 32 per cent, and Kirwan 24.' *Addit. Obſerv.* pag. 120.

'In proportion as folutions are more diluted, their lofs in evaporation at equal temperatures is greater.

DIVISION III.

Decomposition of NITRATE of AMMONIAC : preparation of RESPIRABLE NITROUS OXIDE ; its ANALYSIS.

———

I. *Of the heat required for the decomposition of NITRATE of AMMONIAC.*

THE decompofition of nitrate of ammoniac has been fuppofed by Cornette* to take place at temperatures below 212°, and its fublimation at 234°.

Kirwan, from the non-coincidence in the accounts of its compofition, has imagined that it is partially decompofable, even by a heat of 80°.†

To afcertain the changes effected by increafe of temperature in this falt, a glafs retort was provided, tubulated for the purpofe of introducing

———

* Mem. Par. 1783. See Irifh Tranf. vol. 4.

† Addit. Obf. pag. 120.

the bulb of a thermometer. After it had been made to communicate with the mercurial air-holder, and placed in a furnace, the heat of which could be easily regulated, the thermometer was introduced, and the retort filled with the salt, and carefully luted; so that the appearances produced by different temperatures could be accurately observed, and the products evolved obtained.

From a number of experiments made in this manner on different salts, the following conclusions were drawn.

1st. Compact, or dry nitrate of ammoniac, undergoes little or no change at temperatures below 260°.

2dly. At temperatures between 275° and 300°, it slowly sublimes, without decompofition, or without becoming fluid.

3dly. At 320° it becomes fluid, decompofes, and still slowly sublimes; it neither affuming, or continuing in, the fluid state, without decompofition.

4thly. At temperatures between 340° and 480°, it decompofes rapidly.

5thly. The prifmatic and fibrous nitrates of ammoniac become fluid at temperatures below 300°, and undergo ebullition at temperatures between 360° and 400°, without decompofition.

6thly. They are capable of being heated to 430° without decompofition, or fublimation, till a certain quantity of their water is evaporated.

7thly. At temperatures above 450° they undergo decompofition, without previoufly lofing their water of cryftalifation.

II. *Decompofition of Nitrate of Ammoniac ; production of refpirable Nitrous Oxide ; its properties.*

200 grains of compact nitrate of ammoniac were introduced into a glafs retort, and decompofed flowly by the heat of a fpirit lamp. The firft portions of the gas that came over were rejected, and the laft received in jars containing

mercury. No luminous appearance was perceived in the retort during the procefs, and almoft the whole of the falt was refolved into fluid and gas. The fluid had a faint acid tafte, and contained fome undecompounded nitrate. The gas collected exhibited the following properties.—

a. A candle burnt in it with a brilliant flame, and crackling noife. Before its extinction, the white inner flame became furrounded with an exterior blue one.

b. Phofphorus-introduced into it in a ftate of inflammation, burnt with infinitely greater vividnefs than before.

c. Sulphur introduced into it when burning with a feeble blue flame, was inftantly extinguifhed; but when in a ftate of active inflammation (that is, forming fulphuric acid) it burnt with a beautiful and vivid rofe-colored flame.

d. Inflamed charcoal, deprived of hydrogene, introduced into it, burnt with much greater vividnefs than in the atmofphere.

e. To fome fine twifted iron wire a fmall piece of cork was affixed : this was inflamed, and the whole introduced into a jar of the air. The iron burned with great vividnefs, and threw out bright fparks as in oxygene.

f. 30 meafures of it expofed to water pre-vioufly boiled, was rapidly abforbed ; when the diminution was complete, rather more than a meafure remained.

g. Pure water faturated with it, gave it out again on ebullition, and the gas thus produced retained all its former properties.

h. It was abforbed by red cabbage juice ; but no alteration of color took place.

i. Its tafte was diftinctly fweet, and its odor flight, but agreeable.

j. It underwent no diminution when min-gled with oxygene or nitrous gas.

Such were the obvious properties of the NITROUS OXIDE, or the gas produced by the decompofition of nitrate of ammoniac in a tem-perature not exceeding 440°. Other properties of it will be hereafter demonſtrated, and its affinities fully inveſtigated.

III. *Of the gas remaining after the abſorption of Nitrous Oxide by Water.*

In expoſing nitrous oxide at different times to rain or ſpring water, and water that had been lately boiled, I found that the gas remaining after the abſorption was always leaſt when boiled water was employed, though from the mode of production of the nitrous oxide, I had reaſon to believe that its compoſition was generally the ſame.

This circumſtance induced me to ſuppoſe that ſome of the reſiduum might be gas previouſly contained in the water, and liberated from it in conſequence of the ſtronger affinity of that fluid for nitrous oxide. But the greater part of it, I conjectured to conſiſt of nitrogene produced in conſequence of a complete decompoſition of part of the acid, by the hydrogene. It was in endeavoring to aſcertain the relative purity of nitrous oxide produced at different periods of the procefs of the decompofition of nitrate of ammoniac, 'that I difcovered the true reafon of the appearance of refidual gas.

I decompofed fome pure nitrate of ammoniac in a fmall glafs retort ; and after fuffering the firft portions to efcape with the common air, I caught the remainder in three feparate veffels ftanding in the fame trough, filled with water that had been long boiled, and which at the time of the experiment was fo warm that I could fcarcely bear my hands in it. The different quantities collected gave the fame intenfe brilliancy to the flame of a taper.

26 meafures of each of them were feparately inferted into 3 graduated cylinders, of nearly the fame capacity, over the fame boiled water. As the water cooled, the gas was abforbed by agitation. When the diminution was complete, the refiduum in each cylinder filled, as nearly as poffible, the fame fpace ; about two thirds of a meafure.

To each of the refiduums I added two mea-

fures of nitrous gas; they gave copious red
vapor, and after the condenfation filled a fpace
rather lefs than two meafures.

I Hence the refidual gas contained more
oxygene than common air.

I now introduced 26 meafures of gas from
one of the veffels into a cylinder filled with
unboiled fpring water of the fame kind.*. After
the abforption was complete, near two meafures
remained. Thefe added to two meafures of
nitrous air, diminifhed to 2,5 nearly.

Thefe experiments induced me to believe
that the refidual gas was not produced in the
decompofition of nitrate of ammoniac, but that
it was wholly liberated from the water.

To afcertain this point with precifion, I
diftilled a fmall quantity of the fame kind of
water, which had been near an hour in ebullition, into a graduated cylinder containing
mercury. To this I introduced about one third

---

* Two meafures of air difpelled from this water by
boiling, mingled with 2 of nitrous gas, diminifhed to 2,4
nearly.

of its bulk, i. e. 12 meafures of nitrous oxide, which had been carefully generated in the mercurial apparatus. . After the abforption, a fmall globule of gas only remained, which could hardly have equalled one fourth of a meafure. On admitting to this globule a minute quantity of nitrous gas, an evident diminution took place.

Though this experiment proved that in proportion as the water was free from air, the refiduum was lefs, and though there was no reafon to fuppofe that the ebullition and diftillation had freed the water from the whole of the air it had held in folution, ftill I confidered a decifive experiment wanting to determine whether nitrous oxide was the only gas produced in the flow decompofition of nitrate of ammoniac, or whether a minute quantity of oxygene was not likewife evolved.

I received the middle part of the product of a decompofition of nitrate of ammoniac, under a cylinder filled with dry mercury, and introduced to it fome ftrong folution of ammoniac. After the white cloud produced by the combination of the ammoniacal vapor with the nitric acid, fufpended in the nitrous oxide, had been completely precipitated, I introduced a fmall quantity of nitrous gas. No white vapor was produced.

Now if any gas combinable with nitrous gas had exifted in the cylinder, the quantity of nitrous acid produced, however fmall, would have been rendered perceptible by the ammoniacal fumes; for when a minute globule of common air was admitted into the cylinder, white clouds were inftantly perceptible.

It feems therefore reafonable to conclude,

1. That the refidual gas of nitrous oxide, is air previoufly contained in the water, (which in no cafe can be perfectly freed from it by ebullition), and liberated by the ftronger attraction of that fluid for nitrous oxide.

2. That nitrate of ammoniac, at temperatures below 440°, is decompounded into pure nitrous oxide, and fluid:

3. That in afcertaining the purity of nitrous oxide from its abforption by water, corrections ought to be made for the quantity of gas dif-

pelled from the water. This quantity in common water diftilled under mercury, being about $\frac{1}{50}$; in water fimply boiled, and ufed when hot, about $\frac{1}{30}$; and in common fpring water, $1\frac{1}{12}$.

IV. *Specific gravity of Nitrous Oxide.*

To underftand accurately the changes taking place during the decompofition of nitrate of ammoniac, we muft be acquainted with the fpecific gravity and compofition of nitrous oxide.

90 cubic inches of it, containing about $\frac{1}{35}$ common air, introduced from the mercurial airholder into an exhaufted globe, increafed it in weight 44,75 grains; thermometer being 51°, and atmofpheric preffure 30,7.

106 cubic inches, of fimilar compofition, weighed in like manner, gave at the fame temperature and preffure nearly 52,25 grains; and in another experiment, when the thermometer was 41°, 53 grains.

So that accounting for the fmall quantity of

common air contained in the gafes weighed; we
may conclude, that 100 cubic inches of pure
nitrous oxide weigh 50,1 grains at temperature
50°, and atmofpheric preffure 37.

I was a little furprifed at this great fpecific
gravity, particularly as I had expected, from
Dr. Prieftley's obfervations, to find it lefs heavy
than atmofpherical air. This philofopher fuppofed, from fome appearances produced by the
mixture of it with aëriform- ammoniac, that it
was even of lefs fpecific gravity than that gas.[*]

V. *Analyfis of Nitrous Oxide.*

The nitrous oxide may be analifed, either by
charcoal or hydrogene ; during the combuftion
of other bodies in it, fmall portions of nitrous
acid are generally formed, as will be fully explained hereafter.

The gas that I employed was generated from

---

[*] Experiments and Obfervations, vol. 2, pag. 89. Laft
Edition.

compact nitrate of ammoniac, and was in its highest state of purity, as it left a residuum of 38 only, when absorbed by boiled water.

10 cubic inches of it were inserted, into a jar graduated to 1 cubic inches, containing dry mercury. Through this mercury a piece of charcoal which had been deprived of its hydrogene by long exposure to heat, weighing about a grain, was introduced, while yet warm: No perceptible absorption of the gas took place.*

Thermometer being 46°, the focus of a lens was thrown on the charcoal, which instantly took fire, and burnt vividly for about a minute, the gas being increased in volume. After the vivid combustion had ceased, the focus was again thrown on the charcoal; it continued to burn for near ten minutes, when the process stopped.

The gas, when the original pressure and temperature were restored, filled a space equal to 12,5 cubic inches.

---

* A minute quantity, however, must have been absorbed, and given out again when the charcoal was heated.

On introducing to it a fmall quantity of ftrong folution of ammoniac*, white vapor was inftantly perceived, and after a fhort time the reduction was to about 10,1 cubic inches; fo that apparently, 2,4 cubic inches of carbonic acid had been formed. The 10,1 cubic inches of gas remaining were expofed to water which had been long in ebullition, and which was introduced whilft boiling, under mercury. After the abforption of the nitrous oxide by the water, the gas remaining was equal to 5,3.

But on combining a cubic inch of pure nitrous oxide with fome of the fame water, which had been received under mercury in a feparate veffel, nearly $\frac{1}{22}$ remained. Confequently we may conclude, that 5,1 of a gas unabforbable by water, was produced in the combuftion.

This gas extinguifhed flame, gave no diminution with oxygene, and the flighteft poffible

* Strong folution of ammoniac has no attraction for nitrous oxide.

G

with nitrous gas. When an electric spark was passed through it, mingled with oxygene; no inflammation, or *perceptible* diminution took place.✝ We may consequently conclude that it was nitrogene, mingled with a minute portion of common air, expelled from the water.

The charcoal was diminished in bulk to one half nearly, but the loss of weight could not be ascertained, as its pores were filled with mercury.

Now 5 cubic inches of nitrous oxide were absorbed by the water, consequently 5 were decompounded by the charcoal; and these produced 5,1 cubic inches of nitrogene; and by giving their oxygene to the charcoal, apparently 2,4 of carbonic acid.

But 5 cubic inches of nitrous oxide weigh 2,5 grains, and 5,1 cubic inches of nitrogene 1,55; then 2,5 — 1,55 = ,95.

So that reasoning from the relative specific

| The gas was examined by those tests in order to prove that no water had been decomposed.

gravities of nitrogene and nitrous oxide, 2,5 grains of the laft are compofed of 1,55 nitrogene, and ,95 oxygene.

But from many experiments made on the fpecific gravity of carbonic acid, in Auguft, 1799, I concluded that 100 cubic inches of it weighed 47,5 grains, thermometer being 60,1°, and barometer 29,5. Confequently, making the neceffary corrections, 2,4 cubic inches of it weigh 1,14 grains; and on Lavoifier's and Guyton's * eftimation of its compofition, thefe 1,13 grains contain 8,2 of oxygene.

So that, drawing conclufions from the quantity of carbonic acid formed in this experiment, 2,5 grains of nitrous oxide will be compofed of 82 oxygene, and 1,68 nitrogene.

The difference between thefe eftimations is confiderable, and yet not more than might have been expected, if we confider the probable fources of error in the experiment.

---

* See the curious paper of this excellent philofopher, on the combuftion of the 'diamond, in which he proves that charcoal is, in fact, oxide of diamond. Annales de Chimie. xxxi.

· 1. It is likely that variable minute quantities of hydrogene remain combined with charcoal, even after it has been long expofed to a red heat.

· 2. It is probable that the nitrogene and carbonic acid produced were capable of diffolving more water than that held in folution by the nitrous oxide; and if fo, they were more condenfed than if faturated with moifture, and hence the quantity of carbonic acid under-rated.

We may confequently fuppofe the eftimation founded on the quantity of nitrogene evolved, moft correct; and making a fmall allowance for the difference, conclude, that 100 grains of nitrous oxide are compofed of about 37 oxygene, and 63 nitrogene; exifting in a much more condenfed ftate than when in their fimple forms.

The tolerable accuracy of this ftatement will be hereafter demonftrated by a number of experiments on the combuftion of different bodies in nitrous oxide, detailed in Refearch II.

VI. *Minute examination of the decompofition of Nitrate of Ammoniac.*

Into a retort weighing 413,75 grains, and of the capacity of 7,5 cubic inches, 100 grains of pulverifed compact nitrate of ammoniac were introduced. To the neck of this retort was adapted a recipient, weighing 711 grains, tubulated for the 'purpofe of communicating with the mercurial airholder, and of the capacity of 8,3 cubic inches.

Temperature being 50°, and atmofpheric preffure 30,6, the recipient was inferted into a veffel of cold water, and made to communicate with the airholder. The heat of a fpirit lamp was then flowly applied to the retort : the falt quickly began to decompofe, and to liquify. The temperature was fo regulated, as to keep up an equable and flow decompofition.

During this decompofition, no luminous appearance was perceived in the retort; the gas that came into the airholder was very little clouded, and much water condenfed in the receiver.

After the procefs was finifhed, the communication between the mercurial airholder and the recipient was preferved till the common temperature was reftored to the retort.

The volume of the gas in the cylinder was 85,5 cubic inches. The abfolute quantity of nitrous oxide in thofe 85,5 cubic inches, it was difficult to afcertain with great nicety, on account of the common air previoufly contained in the veffels.

45 meafures of it, expofed to well boiled water, diminifhed by agitation to 8 meafures. So that reafoning from the quantity of air, which fhould have been expelled from the water by the nitrous oxide, we may conclude that the 85,5 cubic inches were nearly pure.

The retort now weighed 419,25 grains, confequently 5,5 grains of falt remained in it. This falt was chiefly collected about the lower part of the neck, and contained rather more water than the compact nitrate, as in fome places it was cryftalifed.

The recipient with the fluid it contained, weighed 759 grains. It had confequently gained in weight 48 grains.

Now the 85,5 cubic inches of nitrous oxide produced, weigh about 42,5 grains; and this added to 48 and 5,5, $= 96$ grains; fo that about 4 grains of falt and fluid were loft, probably by being carried over and depofited by the gas.*

As much of the fluid as could be taken out of the recipient, weighed 46 grains, and held in folution much nitrate of ammoniac with fuperabundance of acid. This acid required for its faturation, $3\frac{1}{8}$ of carbonate of ammoniac (containing, as well as I could guefs), about 20 per cent alkali.

The whole folution evaporated, gave 18 grains of compact nitrate of ammoniac. But

---

* This was actually the cafe; for on examining the conducting tube the day after the experiment, fome minute cryftals of prifmatic nitrate of ammoniac were perceived in it.

reafoning from the quantity of carbonate of
ammoniac employed, the free nitric acid was
equal to 2,75 grains, and this muft have formed
3,56 grains of falt. Confequently the falt pre-
exifting in the folution was about 14,44 grains.

But befides the fluid taken out of the recipient,
2 grains remained in it : let us fuppofe this,
and the 4 grains loft, to contain 2 of falt, and
,6 of free acid.

Then the undecompounded

falt is 5,5 + 14,4 + 2	= 21,9
The free acid 2,75 + ,6	= 3,35
Gas - - -	42,5
Water - -	32,25
	———
	100

Now about 78,1 grains of falt were decompounded, and formed into 42.5 grains of gas,
3,35 grains acid, and 32,25 grains water.

But there is every reafon to fuppofe, that in
this procefs, when the hydrogene of the ammoniac combines with a portion of the oxygene of
the nitric acid to form water, and the nitrogene

enters into union with the nitrogene and remaining oxygene of the nitric acid, to form
nitrous oxide; that water pre-exifting in nitric
acid and ammoniac, fuch as they exifted in
the aëriform ftate, is depofited with the water
produced by the new arrangement, and not
wholly combined with the nitrous oxide formed.
Hence it is impoffible to determine with great
exactitude, the quantity of water which was
abfolutely formed in this experiment.

78,1 grains of falt are compofed of 15,4
alkali, 58 acid, and 4,7 water.

And reafoning from the different affinities of
water for nitric acid, ammoniac, and nitrous
oxide, it is probable that ammoniac, in its decompofition, divides its water in fuch a ratio,
between the nitrogene furnifhed to the nitrous
oxide, and the hydrogene entering into union
with the oxygene of the nitric acid, as to enable
us to affume, that the hydrogene requires for
its faturation nearly the fame quantity of oxygene as when in the aëriform ftate; or that it
certainly cannot require lefs.

., But 15,4 alkali contain 3,08 hydrogene, and 12,32 nitrogene ;* and 3,08 hydrogene require 17,4 of oxygene to form 20,48 of water.

Now 32,5 grains of water exifted before the experiment ; 4,7 grains of water were contained by the falt decompofed, and 32,5 — 4,7 = 27,8 : and 27,8 — 20,48, the quantity generated, = 7,52, the quantity exifting in the nitric acid.

But the nitric acid decompofed is 58^g — 3,35 = to 54,7 ; and 54,7 — 7,5 = 47,2, which entered into new combinations. Thefe 47,2 confift of 33,2 oxygene, and 14, nitrogene. And 33,2 —17,4, the quantity employed to form the water, = 15,8, which combined with 14,, nitrogene of the nitric acid, and 12,32 of that of the ammoniac, to form 42,12 of nitrous oxide. And on this eftimation, 100 parts of nitrous oxide would contain 37,6 oxygene, and 62,4 nitrogene ; a computation much nearer the refults of the analyfis than could

* Owing part of their weight to an unknown quantity of water.

have been expected, particularly as fo many unavoidable fources of error exifted in the procefs.

The experiment that I have detailed is the moft accurate of four, made on the fame quantity of falt. The others were carried on at rather higher temperatures, in confequence of which, more water and falt were fublimed with the gas.

To Berthollet, we owe the difcovery of the products evolved during the flow decompofition of nitrate of ammoniac; but as this philofopher in his examination of this procefs, chiefly defigned to prove the exiftence of hydrogene in ammoniac, he did not afcertain the quantity of gas produced, or minutely examine its properties; from two of them, its abforption by water and its capability of fupporting the vivid combuftion of a taper; he inferred its identity with the dephlogifticated nitrous gas of Prieftley, and concluded that it was nitrous gas with excefs of pure air.*

* Mem. de Paris. 1785, and Journal de Phyfique, 1786, page 175.

VII. *Of the heat produced during the decom-position of nitrate of ammoniac.*

To afcertain whether the temperature of nitrate of ammoniac was increafed or diminifhed after it had been raifed to the point effential to its decompofition, during the evolution of ni-trous oxide and water; that is, in common lan-guage, whether heat was generated or abforbed in the procefs; I introduced a thermometer into about 1500 grains of fibrous nitrate of ammo-niac, rendered liquid in a deep porcelain cup. During the whole of the evaporation, the tem-perature was about 380°, the fire being care-fully regulated.

As foon as the decompofition took place, the thermometer began to rife; in lefs than a quar-ter of a minute it was 410°, in two minutes it was 460°.

The cup was removed from the fire; the de-compofition ftill went on rapidly, and for about a minute the thermometer was ftationary. It then gradually and flowly fell; in three minutes it was 440°, in five minutes 420°, in feven minutes 405°, in nine minutes 360°, and in thirteen minutes 307°, when the decompofition had nearly ceafed, and the falt began to folidify.

From this experiment, it is evident that an increafe of temperature is produced by the decompofition of nitrate of ammoniac : though the capacity of water and nitrous oxide for heat, fuppofing the truth of the common doctrine, and reafoning from analogy, muft be confiderably greater than that of the falt.

VIII. *Of the decompofition of Nitrate of Ammoniac at high temperatures, and production of Nitrous gas, Nitrogene, Nitrous Acid, and Water.*

At an early period of my inveftigation relating to the nitrous oxide, I difcovered that when a heat above 600° was applied to nitrate of ammoniac, fo that a vivid luminous appearance was produced in the retort, certain portions of nitrous gas, and nitrogene, were evolved with the

nitrous oxide. But I was for fome time ignorant of the precife nature of this decompofition,
and doubtful with regard to the poffibility of
effecting it in fuch a manner as to prevent the
production of nitrous oxide altogether.

I firft attempted to decompofe nitrate of
ammoniac at high temperatures, by introducing
it into a well coated green glafs retort, having
a wide neck, communicating with the pneumatic apparatus, and ftrongly heated in an air-
furnace. But though in this procefs a detonation always took place, and much light was produced, yet ftill the greater portion of the gas
generated was nitrous oxide ; the nitrous gas
and nitrogene never amounting to more than
one third of the whole.

After breaking many retorts by explofions,
without gaining any accurate refults, I employed a porcelain tube, curved fo as to be
capable of introduction into the pneumatic
apparatus, and clofed at one end.

The clofed end was heated red, nitrate of
ammoniac introduced into it, and all the latter

portions of gas produced in the explofion, received in the pneumatic apparatus, filled with
warm water.

Three explofions were required to fill a jar
of the capacity of 20 cubic inches. The gas
produced in the firft, when it came over, was
tranfparent and dark orange, fimilar in its
appearance to the nitrous acid gas produced in
the firft experiment; but it fpeedily became
white and clouded, whilft a flight diminution
of volume took place.

When the fecond portion was generated and
mingled with the clouded gas, it again became
tranfparent and yellow for a fhort time, and then
affumed the fame appearance as before.

The water in the trough, after this experiment, had an acid tafte, and quickly reddened cabbage juice rendered green by an
alkali

6 cubic inches of the gas produced were
expofed to boiled water, but little or no abforption took place. Hence, evidently, it contained no nitrous oxide.

They were then expofed to folution of fulphate of iron: the folution quickly became dark-
colored, and an abforption of 1,6 took place
on agitation.*

The gas remaining inftantly extinguifhed the
taper, and was confequently nitrogene.

This experiment was repeated, with nearly
the fame refults.

We may then conclude, that at high temperatures, nitrate of ammoniac is wholly refolved into water, nitrous acid, nitrous gas,
and nitrogene ; whilft a vivid luminous appearance is produced.

The tranfparency and orange color produced
in the gas that had been clouded, by new portions of it, doubtlefs arofe from the folution of
the nitric acid and water forming the cloud, in
the heated nitrous vapor produced, fo as to conftitute an aëriform triple compound ; whilft the
cloudinefs and abforption fubfequent were pro-

---

* The abforption of nitrous gas by fulphate of iron, &c.
will be treated of in the next divifion.

duced by the diminifhed temperature, which deftroyed the ternary combination, and feparated the nitrous acid and water from the nitrous gas,

From the rapidity with which the deflagration of nitrate of ammoniac proceeds, and from the immenfe quantity of light produced, it is reafonable to fuppofe that a very great increafe of temperature takes place. The tube in which the decompofition has been effected, is always ignited after the procefs.

IX. *Speculations on the decompofitions of Nitrate of Ammoniac.*

All the phænomena of chemiftry concur in proving, that the affinity of one body, A, for another, B, is not deftroyed by its combination with a third, C, but only modified ; either by condenfation, or expanfion, or by the attraction of C for B.

On this principle, the attraction of compound bodies for each other muft be revolved into the reciprocal attractions of their conftituents, and confequently the changes produced in them by variations of temperature explained, from the alterations produced in the attractions of thofe conftituents.

Thus in nitrate of ammoniac, four affinities may be fuppofed to exift :

1. That of hydrogene for nitrogene, produ- cing ammoniac.

2. That of oxygene for nitrous gas, producing nitric acid.

3. That of the hydrogene of ammoniac for the oxygene of nitric acid.

4. That of the nitrogene of ammoniac for the nitrous gas of nitric acid.

At temperatures below 300°, the falt, from the equilibrium between thefe affinities, pre- ferves its exiftence.

Now when its temperature is raifed to 400°, the attractions of hydrogene for nitrogene,* and

---

* As is evident from the decompofition of ammoniac by heat.

of nitrous gas for oxygene,‡ are diminished; whilft the attraction of hydrogene for oxygene† is increafed; and perhaps that of nitrogene for nitrous gas.

Hence the former equilibrium of affinity is deftroyed, and a new one produced.

The hydrogene of the ammoniac combines with the oxygene of the nitric acid to generate water; and the nitrogene of the ammoniac enters into combination with the nitrous gas to form nitrous oxide: and the water and nitrous oxide produced, moft probably exift in binary combination in the aëriform ftate, at the temperature of the decompofition.

But when a heat above 800° is applied to nitrate of ammoniac, the attractions of nitrogene and hydrogene for each other, and of

---

‡ Nitric acid is phlogifticated by heat, as appears from Dr. Prieftley's experiments. Vol. 3, p. 26.

† As is evident from the increafe of temperature required for the formation of water.

oxygene for nitrous gas,* are, ftill more diminifhed ; whilft that of nitrogene for nitrous gas
is deftroyed, and, that of hydrogene for oxygene increafed to a great extent : likewife
a new attraction takes place; that of nitrous
gas for nitric acid, to form nitrous vapor.†
Hence a new arrangement of principles is
rapidly produced ; the nitrogene of ammoniac

* For ammoniac and nitrous oxide are both decompofed at
the red heat, and oxygene given out from nitric acid when
it is paffed through a, heated tube.

† Whenever nitrous acid is produced at high temperatures, it is always highly phlogifticated, provided it has not
been long in contact with oxygene. When Dr. Prieftley
paffed nitric acid through a tube heated red, he procured
much oxygene, and phlogifticated acid ; and the water
in the apparatus employed was fully impregnated with
nitrous air. Hence it would appear, that heat diminifhes
the attraction between oxygene and nitrous gas; and increafes the affinity of nitrous gas for nitrous acid. Mr.
JAMES THOMSON, whofe theory of the Nitrous Acid I have
already mentioned, from fome experiments on the phlo-
giftication of Nitric Acid by heat, which he has communicated to me, concludes with great juftnefs, that a portion
of the acid is always completely decompofed in this procefs:
the oxygene liberated, and the nitrous gas combined with
the remaining acid.

having no affinity for any of the fingle principles
at this temperature, enters into no binary compound : the oxygene of the nitric acid forms
water with the hydrogene, and the nitrous gas
combines with the nitric acid to form nitrous
vapor. All thefe fubftances moft probably
exift in combination at the temperature of their
production ; and at a lower temperature, affume
the forms of nitrous acid, nitrous gas, nitrogene,
and water.

I have avoided entering into any difcuffions
concerning the light and heat produced in this
procefs ; becaufe thefe phænomena cannot be
reafoned upon as ifolated facts, and their relation
to general theory will be treated of hereafter.

X. *On the preparation of Nitrous Oxide for
experiments on Refpiration.*

When compact nitrate of ammoniac is flowly
decompofed, the nitrous oxide produced is
almoft immediately fit for refpiration ; but as
one part of the falt begins to decompofe before

the other is rendered fluid, a confiderable lofs is produced by fublimation.

For the production of large quantities of nitrous oxide, fibrous nitrate of ammoniac fhould be employed. This falt undergoes no decompofition till the greater part of its water is evaporated, and in confequence at the commencement of that procefs, is uniformly heated.

The gas produced from fibrous nitrate, muft be fuffered to reft at leaft for an hour after its generation. At the end of this time it is generally fit for refpiration. If examined before, it will be found to contain more or lefs of a white vapor, which has a difagreeable acidulous tafte, and ftrongly irritates the fauces and lungs. This vapor, moft probably, confifts of acid nitrate of ammoniac and water, which were diffolved by the gas at the temperature of its production, and afterwards flowly precipitated.

It is found in lefs quantity when compact nitrate is employed, becaufe more falt is fublimed in this procefs, which being rapidly precipitated, carries with it the acid and water.

Whatever salt is employed, the last portions of gas produced, generally contain less vapor, and may in consequence be respired sooner than the first.

The nitrate of ammoniac should never be decomposed in a metallic vessel,* nor the gas produced suffered to come in contact with any metallic surface; for in this case the free nitric acid will be decomposed, and in consequence, a certain quantity of nitrous gas produced.

The apparatus that has been generally employed in the medical pneumatic institution, for the production of nitrous oxide, consists

1. Of a glass retort, of the capacity of two or three quarts, orificed at the top, and furnished with a ground stopper.

2. Of a glass tube, conical for the purpose of receiving the neck of the retort; about 4 inches wide in the narrowest part, 4 feet long, curved at the extremity, so as to be capable of

* Except it be gold or platina.

introduction into an airholder, and inclofed by
tin plate to preferve it from injury.

3. Of airholders of Mr. Watt's invention,
filled with water faturated with nitrous oxide.

4. Of a common air-furnace, provided with
dampers for the regulation of the heat.

The retort, after the infertion of the falt, is
connected with the tube, carefully luted, and
expofed to the heat of the furnace, on a con-
venient ftand. The temperature is never fuffered to be above 500°. After the decompofition has proceeded for about a minute, fo that
the gas evolved from the tube enlarges the
flame of a taper, the curved end is inferted
into the airholder, and the nitrous oxide preferved.

The water thrown out of the airholders in
confequence of the introduction of the gas,
is preferved in a veffel adapted for the purpofe, and employed to fill them again; for if
common water was to be employed in every experiment, a great lofs of gas would be produced
from abforption.

A pound of fibrous nitrate of ammoniac, de-compofed at a heat not above 500°, produces nearly 5 cubic feet of gas; whilft from a pound of compact nitrate of ammoniac, rarely more than 4,25 cubic feet can be collected.

For the production of nitrous oxide in quantities not exceeding ·20 quarts, a mode ftill more fimple than that I have juft defcribed may be employed. The falt may be decompofed by the heat of an argands lamp, or a common fire, in a tubulated glafs retort, of 20 or 30. cubic inches in capacity, furnifhed with a long neck, curved at the extremity; and the gas received in fmall airholders.

Thus, if the pleafurable effects, or medical properties of the nitrous oxide, fhould ever make it an article of general requeft, it may be procured with much lefs time, labor, and expence,* than moft of the luxuries, or even neceffaries, of life.

* A pound of nitrate of ammoniac cofts about 5s. 10d. This pound, properly decompofed, produces rather more than 34 moderate dofes of air; fo that the expence of a dofe is about 2d. What fluid ftimulus can be procured at fo cheap a rate?

DIVISION IV.

EXPERIMENTS and OBSERVATIONS on the COMPOSITION of NITROUS GAS, and on its ABSORPTION by different bodies.

I. *Preliminaries.*

IN my account of the composition of nitric acid, in Division I. I gave an estimation of the quantities of oxygene and nitrogene combined in nitrous gas: I shall now detail the experiments on which that estimation is founded.

At an early period of my researches relating to nitrous oxide, from the observation of the phænomena taking place during the production of this substance, I had concluded, that the common opinion with regard to the composition of nitrous gas, was very distant from the truth. I had indeed analysed nitrous gas, by converting it into nitrous oxide, before I attempted to afcertain its compofition by immediately feparating the conftituent principles from each other: and my firft hopes of the poffibility of effecting this, were derived from Dr. Prieftley's experiments on the combuftion of pyrophorus in nitrous gas, and on the changes effected in it, by heated iron and charcoal.

This great philofopher found, that pyrophorus placed in contact with nitrous gas, burnt with great vividnefs, whilft the gas was diminifhed in volume to about one half, which generally confifted of nitrogene and nitrous oxide.* He likewife found, iron heated by a lens in nitrous gas, increafed in weight, whilft the gas was diminfhed about $\frac{1}{2}$, and converted into nitrogene.*

He heated common charcoal, and charcoal of copper,‡ in nitrous gas by a lens. When

---

* Experiments and Obfervations, vol. ii. pag. 50. Laft Edition.

‡ That is, charcoal produced by the decompofition of fpirits of wine. Vol. ii. pag. 39.

common charcoal was employed, the gas was neither increafed or diminifhed in bulk, but wholly converted into nitrogene ; when charcoal of copper was ufed, the volume was a little increafed, and the gas remaining confifted of $\frac{5}{7}$ nitrogene, and $\frac{2}{7}$ carbonic acid.

In his experiments on the iron and pyrophyrus, the nitrous gas was evidently decompofed. From the great quantity of nitrogene produced in thofe on the charcoal, it feems likely that both the common charcoal,* and the charcoal of copper employed contained atmofpherical air, which being difpelled by the heat of the lens,

* Dr. Prieftley fays, " having heated iron in nitrous air, " I proceeded to heat in the fame air, a piece of charcoal " not long after it had been fubjected to a ftrong heat covered " with fand. : The fun not fhining immediately, after the " charcoal was introduced into the veffel of air, through the " mercury by which it was confined, part of the air was " abforbed ; but on heating the charcoal, the quantity was " increafed. Having continued the progrefs as long as I " thought neceffary, I examined the air and found it to be " about as much as the original quantity of nitrous air ; " but it was all phlogifticated air extinguifhing a candle " and having no mixture of fixed air in it."—Experiments and Obfervations, Vol. II, page 39.

was decompofed by the nitrous gas : indeed, till I made the following experiment, I fufpected that the carbonic acid produced, when the charcoal of copper was employed, arofe from a decompofition of the nitrous acid, formed in this way.

I introduced a piece of well-burnt charcoal, which could hardly have weighed the eighth of a grain, whilft red hot, under a cylinder filled with mercury, and admitted to it half a cubic inch of nitrous gas. A flight abforption took place.

The fun being very bright, I kept the charcoal in the focus of a fmall lens for near a quarter of an hour. At the end of this time the gas occupied a fpace nearly as before the experiment, and a very minute portion of the charcoal had been confumed. On introducing into the cylinder a fmall quantity of folution of ftrontian, a white precipitation was perceived, and the gas flowly diminifhed to about three tenths of a cubic inch. To thefe three tenths a little common air was admitted; when very flight red fumes were perceived.

This experiment convinced me, that the attraction of charcoal for the oxygene of nitrous gas, at high temperatures, was fufficiently ftrong to effect a flow decompofition of it.

To be more accurately acquainted with this decompofition, and to learn the quantities of carbonic acid and nitrogene produced from a known quantity of nitrous gas, I proceeded in the following manner.

II. *Analyfis of Nitrous Gas by Charcoal.*

A quantity of nitrous gas was procured in a water apparatus, from the decompofition of nitrous acid by mercury. A portion of it was transferred to the mercurial trough. After the mercury and the jar had been dried by bibulous paper, 40 meafures of this portion were agitated in a folution of fulphate of iron. The gas remaining after the abforption was complete, filled about a measure and half; so that the nitrous gas contained nearly $\frac{1}{26}$ nitrogene.

Thermometer being 53°, a small piece of well burnt charcoal, the weight of which could hardly have equalled a quarter of a grain, was introduced ignited, into a small cylinder filled with mercury, graduated to 10 grain measures; to this, 16 measures, equal to 160 grain m. of nitrous gas, were admitted. An absorption of about one measure and half took place. When the focus of a lens was thrown on the charcoal, a slight increase of the gas was produced, from the emission of that which had been absorbed.

After the process had been carried on for about a half an hour, the charcoal evidently began to fume, and to consume very slowly, though no alteration in the volume of the gas was observed.

The sun not constantly shining, the progress of the experiment was now and then stopped; but taking the whole time, the focus could not have been applied to it for less than four hours. When the process was finished, the gas was increafed in bulk nearly three quarters of a meafure.

A drop of water was introduced into the cylinder, by means of a fmall glafs tube, on the fuppofition that the carbonic acid, and nitrogene, might be capable of holding in folution, more water than that contained in the nitrous gas decompofed; but no alteration of volume took place.

When 20 grain meafures of folution of pale green* fulphate of iron were introduced into the cylinder, they became rather yellower than before, but not dark at the edges, as is always the cafe when nitrous gas is prefent. On agitation, a diminution of nearly half a meafure was produced, doubtlefs from the abforption of fome of the carbonic acid by the folution.

A fmall quantity of cauftic potafh, much more than was fufficient to decompofe the fulphate of iron, was now introduced. A rapid diminution took place, and the gas remaining

---

* That is, fulphate of iron containing oxide of iron, in the firft degree of oxygenation.

filled about 8 meafures. This gas was agitated for fome time over water, but no abforption took place. Two meafures of it were then transferred into a detonating cylinder with two meafures of oxygene. The electric fpark was paffed through them, but no diminution was produced. Hence it was nitrogene, mingled with no afcertainable quantity of hydrogene : confequently little or no water could have been decompofed in the procefs.

Now fuppofing, for the greater eafe of calculation, each of the meafures employed, cubic inches.

16 of nitrous gas — $\frac{1}{26}$ = 15,4 were decompofed, and thefe weigh, making the neceffary corrections, 5,2 ; but 7,4 nitrogene were produced, and thefe weigh about 2,2. So that reafoning from the relative fpecific gravities of nitrous gas and nitrogene, 5,2 grains of nitrous gas will be compofed of 3 oxygene, and 2,2 nitrogene.

But 8,7 of carbonic acid were produced, which weigh 41 grains, and confift of 2,9 oxygene, and 1,2 charcoal.* Confequently, drawing conclufions from the quantity of carbonic acid, formed, 5,2 grains of nitrous gas will confift of 2,9 oxygene, and 2,3 nitrogene.

The difference in thefe eftimations is much lefs than could have been expected ; and taking the mean proportions, it would be inferred from them, that, 100 grains of nitrous gas, contain 56,5 oxygene, and 43,5 nitrogene.

I repeated this experiment with refults not very different, except that the increafe of volume was rather greater, and that more unabforbable gas remained ; which probably depended on the decompofition of a minute quantity of water, that had adhered to the charcoal in paffing through the mercury.

As nitrous gas is decompofable into nitrous acid, and nitrogene, by the electric fpark ; it occurred to me, that a certain quantity of nitrous acid might have been poffibly produced, in the experiments on the decompofition of nitrous gas, by the intenfely ignited charcoal.

* That is, carbon, or oxide of diamond.

To afcertain this circumftance, I introduced into 12 meafures of nitrous gas, a fmall piece of charcoal which had been juft reddened. The fun being very bright, the focus of the lens was kept on it for rather more than an hour and quarter. In the middle of the procefs it began to fume and to fparkle, as if in combuftion. In three quarters of an hour, the gas was increafed rather more than half a meafure; bnt no alteration of volume took place afterwards.

The mercury was not white on the top as is ufually the cafe when nitrous acid is produced. On introducing into the cylinder a little pale green fulphate of iron, and then adding prufiate of potafh, a white precipitate only was produced. Now, if the minuteft quantity of nitric acid had been formed, it would have been decompofed by the pale green oxide of iron, and hence, a vifible quantity of pruffian blue* produced, as will be fully explained hereafter.

* That is, blue pruffiate of iron.

III. *Analysis of Nitrous Gas by Pyrophorus.*

I placed some newly made pyrophorus, about as much as would fill a quarter of a cubic inch, into a jar filled with dry mercury, and introduced to it, four cubic inches of nitrous gas, procured from mercury and nitric acid. It instantly took fire and burnt with great vividness for some moments.

After the combustion had ceased, the gas was diminished about three quarters of a cubic inch. The remainder was not examined; for the diminution appeared to go on for some time, after; in an half hour, when it was compleat, it was to 2 cubic inches. A taper, introduced into these, burnt with an enlarged flame, blue at the edges; from whence it appeared, that they were composed of nitrogene and nitrous oxide.

I now introduced about half a cubic inch of pyrophorus to two cubic inches of nitrous gas; the combustion took place, and the gas was rapidly diminifhed to one half; and on fuffering it to remain five minutes to one-third nearly; which extinguifhed flame.

Sufpecting that this great diminution was owing to the abforption of fome of the nitrogene formed, by the charcoal of the pyrophorus, I carefully made a quantity of pyrophorus; employing more than two-thirds of alumn, to one-third of fugar.

To rather more than half of a cubic inch of this, two cubic inches of nitrous gas, which contained about $\frac{1}{40}$ nitrogene, were admitted. After the combuftion, the gas remaining, *apparently* filled a fpace equal to 1,2 cubic inches; but, as on account of the burnt pyrophyrus in the jar, it was impoffible to afcertain the volume with nicety, it was carefully and wholly transferred into another jar. It filled a fpace equal to 1,15 cubic inches nearly.

When water was admitted to this gas no abforption took place. It underwent no diminution with nitrous gas, and a taper plunged into it was inftantly extinguifhed. We may confequently conclude that it was nitrogene.

Now 2 cubic inches of nitrous gas weigh ,686 grains, and 1,1 of nitrogene —,05, the quantity previoufly contained in the gas = to 1,05, 3,19. Hence ,686 of nitrous gas would be compofed of ,367 oxygene, and ,319 nitrogene; and 100 grains would contain 53,4 oxygene, and 46,6 nitrogene.

IV. *Additional obfervations on the combuftion of bodies in Nitrous Gas, and on its Compofition.*

Though phofphorus may be fufed, and even fublimed, in nitrous gas, without producing the flighteft luminous appearance,* yet when

---

* No luminous appearance is produced when phofphorus is introduced into *pure* nitrous gas. It has been often obferved, that phofphorus is luminous in nitrous gas, that has not been long in contact with water after its production. This phænomenon, I fufpect, depends either on the decompofition of the nitric acid held in folution by the nitrous gas; or on the combination of the phofphorus with oxygene loofely adhering to the binary aëriform compound of nitric acid and nitrous gas. I have not yet examined if nitrous gas can be converted into nitrous oxide by long expofure to heated phofphorus: it appears, however, very probable.

it is introduced into it in a ftate of active inflammation, it burns with almoft as much
vividnefs as in oxygene.* Hence it is evident,
that at the heat of ignition, phofphorus is
capable of attracting the oxygene from the
nitrogene of nitrous gas.

I attempted to analife nitrous gas, by introducing into a known quantity of it, confined
by mercury, phofphorus, in a veffel containing
a minute quantity of oxygene.† The phofphorus was inflamed with an ignited iron wire, by
which, at the moment of the combuftion, the
veffel containing it was raifed from the mercury
into the nitrous gas. But after making in this
way, five of fix unfuccefsful experiments, I
defifted. When the communication between
the veffels was made before the oxygene was
nearly combined with the phofphorus, nitrous

* Perhaps this fact has been noticed before; I have not,
however, met with it in any chemical work.

† This mode of inflaming bodies in gafes, not capable of
fupporting combuftion at low temperatures, will be particularly defcribed hereafter.

acid was formed, which inftantly deftroyed the combuftion ; when, on the contrary, the phof-phorus was fuffered to confume almoft the whole of the oxygene, it was not fufficiently ignited when introduced, to decompofe the nitrous gas.

In one experiment, indeed, the phofphorus burnt for a moment in the nitrous gas; the diminution however was flight, and not more than $\frac{1}{4}$ of it was decompofed.

Sulphur, introduced in a ftate of vivid in-flammation, into nitrous gas, was inftantly extinguifhed.

I paffed a ftrong electric fhock through equal parts of hydrogene and nitrous gas, confined by mercury in a detonating tube ; but no inflam-mation, or perceptible diminution, was pro-duced.

19,2 grain meafures of hydrogene were fired by the electric fhock, with 10 of nitrous oxide, and 6 of nitrous gas ; the diminution was to 17 ; and pale green fulphate of iron admitted to the refiduum, was not difcolored. Confequently the nitrous gas was decompofed by the hydrogene, and as will be hereafter more clearly underftood, nearly as much nitrogene furnifhed by it, as would have been produced from half the quantity of nitrous oxide.

Sufpecting that phofphorated hydrogene might inflame with nitrous gas, I paffed the electric fpark through 1 meafure of phofphorated hydrogene, and 4 of nitrous gas; but no diminution was perceptible. I likewife paffed the electric fpark through 1 of nitrous gas, with 2 of phofphorated hydrogene, without inflammation,

Perhaps if I had tried many other different proportions of the gafes, I fhould have at laft difcovered one, in which they would have inflamed; for, as will be feen hereafter, nitrous oxide cannot be decompofed by the compound combuftible gafes, except definite quantities are employed.

From Dr. Prieftley's experiments on iron and pyrophorus, and from the experiments I have detailed, on charcoal, phofphorus, and hydrogene, it appears that at certain temperatures, nitrous gas is decomposable by moſt of the combuſtible bodies: even the extinction of ſulphur, when introduced into it in a ſtate of inflammation, depends perhaps, on the ſmaller quantity of heat produced by the combuſtion of this body, than that of moſt others.

The analyſis of nitrous gas by charcoal, as affording data for determining immediately the quantities of oxygene and nitrogene, ought to be conſidered as moſt accurate; and correcting it by mean calculations derived from the decompoſition of nitrous gas by pyrophorus and hydrogene, and its converſion into nitrous oxide, a proceſs to be deſcribed hereafter, we may conclude, that 100 grains of nitrous gas are compoſed of 55,95 oxygene, and 44,05 nitrogene; or taking away decimals, of 56 oxygene, and 44 nitrogene.

This eſtimation will agree very well with the mean proportions that would be given from Dr. Prieſtley's experiments on the decompoſition of nitrous gas by iron; but as he never aſcer-

tained the purity of his nitrous gas,* and probably employed different kinds in different experiments, it is impoffible to fix on any one, from
which accurate conclufions can be drawn.

Lavoifier's eftimation of the quantities of oxygene and nitrogene entering into the compofition
of nitrous gas, has been generally adopted.
He fuppofes 64 parts of nitrous gas to be compofed of 43½ of oxygene, and 20½ of nitrogene.†

The difference between this account and
mine is very great indeed; but I have already,
in Divifion 1ft, pointed out fources of error in
the experiments of this great man, on the decompofition of nitre by charcoal; which experiments were fundamental, both to his accounts
of the conftitution of nitrous acid, and nitrous
gas.

---

* Elements Englifh Tranf. edit. i. pag. 216.

† Experiments and Obfervations, Vol. II. pag. 40, 2d. Ed.

V. Of the abforption of Nitrous Gas by Water.

Amongſt the properties of nitrous gas noticed by its great diſcovèrer, is that of abſorbability by water.

In expoſing nitrous air to diſtilled water, Dr. Prieſtley found a diminution of the volume of gas, nearly equal to one tenth of the bulk of the water ; and by boiling the water thus im-pregnated, he procured again a certain portion of the nitrous gas.

Humbolt, in his paper on eudiometry, men-tions the diminution of nitrous gas by water. This diminution, he ſuppoſes to ariſe from the decom-poſition of a portion of the nitrous gas, by the water, and the conſequent formation of nitrate of ammoniac.*

* He ſays, " On a obſervé, (depuis qu'on travaille ſur le
" pureté de l'air) que le gaz nitreux, ſecoué avec l'eau, en
" ſouffre une diminution de volume. Quelques phyſiciens
" attribuent ce changement à une vraie abſorption, à une
" diſſolution du gaz nitreux dans l'eau ; d'autres à l'air con-
" tenu dans les interſtices de tous les fluides. Le cit.
" Vanbreda, à Delft, a fait des recherches très-exaĉtes ſur
" l'influence des eaux de pluie et dè puit, ſur les nombres
" eudiométriques ; et les belles expériences du cit. Haſſen-

· I confefs, that even before the following experiments were made, I was but little inclined
to adopt this opinion : the fmall diminution
of nitrous gas by water, and the uniform limits
of this diminution, rendered it extremely improbable.

a. To afcertain the quantity of nitrous gas

" fratz, fur l'abondance d'oxygène, contenue dans les eaux
" de neige et de pluie, font fuppofer que l'air des interftices
" de l'eau joüe uu rôle important dans l'abforption du gaz
" nitreux. En comparant ces effets avec les phénomènes,
" obfervé dans la decompofition du fulfate de fer, nous fup-
" posâmes, le cit. Taffaert et moi, que le fimple contaĉt du
" gaz nitreux avec l'eau diftillée pourroit bien caufer une
" décompofition de ce dernier. Nous examinâmes foign-
" eufement une petite quantité d'eau diftillée, fecouée avec
" beaucoup de gas nitreux trés-pur, et nous trouvâmes,
" au moyen de la terre calcaire, et l'acide muriatique, qu'il
" s'y forme du *nitrate d'ammoniaque*. L'eau fe décompofe
" en cette opération, par un double affinité de l'oxygene,
" pour le gaz nitreux, et de l'hydrogène pour l'azote ; il fe
" forme de l'acide nitrique et de l'*ammoniaque* ; et, quoique
" la quantité du dernier paroiffe trop petite pour en évaluer
" exaĉtment la quantité, fon exiftence cependant fe mani-,
" fefte, (à ne pas fans douter) par le dégagement des va-
" peurs, qui blanchiffent dans la proximité de l'acide mu-
" riatique. Voilá un fait bien frappant que la compofition
" d'une fubftance alcaline par le contaĉt d'une acide, et de
" l'eau.

Annales de Chimie, t. xxviii. pag. 153.

abforbable by pure water, and the limits of abforption, I introduced, into a glafs retort about
5 ounces of water, which had been previoufly
boiled for fome hours., The neck of the retort,
was inverted in mercury, and the water made
to boil. After a third of it had been diftilled,,
fo that no air could poffibly remain in the retort, the remainder was driven over, and condenfed in an inverted jar filled with mercury.
To three cubic inches of this water,* confined
in a cylinder graduated to ,o5 cubic inches, 5
cubic inches of nitrous gas, containing nearly one
thirtieth nitrogene, were introduced.

After agitation for near an hour, rather more
than $\frac{4}{20}$ of a cubic inch appeared to be abforbed;
but though the procefs was continued for near
two hours longer, no further diminution took
place.

The remaining gas was introduced into a
tube graduated to ,o2 cubic inches. It mea-
fured $\frac{14}{50}$; hence $\frac{11}{30}$ had been abforbed.

---

* Which was certainly as free from air as it ever can be
obtained.

Confequently, 100 cubic inches of pure water are capable of abforbing 11,8 of nitrous gas.
In the water thus impregnated with nitrous
gas I could diftinguifh no peculiar tafte;* it
did not at all alter the color of blue cabbage
juice.

b. To determine if the abforption of nitrous
gas was owing to a decompofition of it by the
water, as Humbolt has fuppofed, or to a fimple
folution; I procured fome nitrous gas from
nitrous acid and mercury, containing about
one feventieth nitrogene. 5 cubic inches of
it, mingled with 25, of oxygene, from fulphuric acid and manganefe left a refiduum of
,03. 5 cubic inches more were introduced
to 3 of water, procured in the fame manner as
in the laft experiment, in the fame cylinder.

* Dr. Prieftley found diftilled water, faturated with
nitrous air, to acquire an aftringent tafte and pungent
fmell. In fome unboiled impregnated pump water, I once
thought that I perceived a fubacid tafte; but it was extremely flight, and probably owing to nitrous acid formed
by the union of the oxygene of the common air in the water, with fome of the nitrous gas.

After the diminution was complete, the cylinder was transferred in a small vessel containing mercury, into a water bath, and nearly covered by the water.

As the bath was heated, small globules of gas were given out from the impregnated water, and when it began to boil, the production of gas was still more rapid. After an hour's ebullition, the volume of heated gas was equal to 1,4 cubic inches nearly.

The cylinder was now taken out of the bath, and quickly rendered cool by being placed in a water apparatus. At the common temperature the gas occupied, as nearly as possible, the space of ,5 cubic inches: these ,5 mingled with ,25 of oxygene, of the same kind as that employed before, left a residuum nearly equal to ,93.

From this experiment, which was repeated with nearly the same results, it is evident,

1, That nitrous gas is not decomposable by pure water.

2, That the diminution of volume of nitrous gas placed in contact with water, is owing to a simple solution of it in that fluid.

. 3, That at the temperature of 212°, nitrous gas is incapable of remaining in combination with water. ·

. Humbolt's opinion relating to the decompofition of nitrous gas by water, is founded upon the difengagement of vapor from diftilled water impregnated with nitrous gas, by means of lime, which became white in the proximity of the muriatic acid. But this is a very imperfect, and fallacious teft, of the prefence of ammoniac. I have this day, April 2, 1800, heated 4 cubic inches of diftilled water, impregnated with nitrous gas, with cauftic lime; the vapor certainly became a little whiter when held over a veffel containing muriatic acid; but the vapor of diftilled water produced precifely the fame appearance,* which was owing, moft likely, to

---

* As carbonic acid and ammoniac are both products of animalifation, is it not probable that our common waters particularly thofe in, and near towns and cities, contain carbonate of ammoniac? If fo, this falt will always exift in them after diftillation. In the experiments on carbonate of ammoniac, to which I have often alluded, I found, in diftilling a folution of this falt in water, that before half of

K

the combination of the acid with the aqueous vapor. Indeed, when I added a particle of nitrate of ammoniac, which might have equalled one twentieth of a grain, to the lime and impregnated water, the increafed whitenefs of the vapor was but barely perceptible, though this quantity of nitrate of ammoniac is much more confiderable than that which could have been formed, even fuppofing the nitrous gas decompofed.

VI. *Of the absorption of Nitrous Gas by Water of different kinds.*

In agitating nitrous gas over fpring water, the diminution rarely amounts to more than one thirtieth, the volume of water being taken as unity. I at firft fufpected that this great dif-

the water had paffed into the recipient, the carbonate of ammoniac had fublimed; fo that the diftilled folution was much ftronger than before, whilft the water remaining in the retort was taftelefs. Will this fuppofition at all explain Humbolt's miftake?

ference in the quantity of gas abforbed by fpring water, and pure water, depended on carbonio acid contained in the laft, diminifhing the attraction of it for nitrous gas : but by long boiling a quantity of fpring water confined by mercury, I obtained from it about one twentieth of its bulk of air, which gave nearly the fame diminution with nitrous gas, as atmofpheric air.

This fact induced me to refer the difference of diminution to the decompofition of the atmofpheric air held in folution by the water, the oxygene of which I fuppofed to be converted into nitric acid, by the nitrous gas, whilft the nitrogene was liberated ; and hence the increafed refiduum.

 a. I expofed to pure water, that is, water procured by diftillation under mercury, nitrous gas, containing a known quantity of nitrogene. After the abforption was complete, I found the fame quantity of nitrogene in the refiduum, as was contained in a volume of gas equal to the whole quantity employed.

b. Spring water boiled for fome hours, and fuffered to cool under mercury, abforbed a quantity of nitrous gas equal to one thirteenth of its bulk; which is not much lefs than that abforbed by pure water.

c. I expofed to fpring water, 10 meafures of nitrous gas; the compofition of which had been accurately afcertained; the diminution was one twenty-cighth, the volume of water being taken as unity. On placing the refiduum in contact with folution of fulphate of iron, the nitrogene remaining was nearly one-twentieth more than had been contained by the gas before its expofure to water.

d. Diftilled water was faturated with common air, by being agitated for fome time in the atmofphere. Nitrous gas placed in contact with this water, underwent a diminution of $\frac{1}{18}$; the volume of water being unity. The gas remaining after the abforption contained about one twenty - feventh nitrogene more than before.

e. Nitrous gas expofed to water combined

with about one fourth of its volume of carbonic acid, diminished to $\frac{1}{32}$* nearly. The remainder contained little or no superabundant nitrogene.

From these observations it appears, that the different degrees of diminution of nitrous gas by different kinds of water, may depend upon various causes.

1. Less nitrous gas will be absorbed by water holding in solution earthy salts, than by pure water; and in this case the diminution of the attraction of water for nitrous gas will probably be in the ratio of the quantities of salt combined with it. *a. b.*

2. The apparent diminution of nitrous gas in water, holding in solution atmospheric air, will be less than in pure water, though the absolute diminution will be greater; for the same portion will be absorbed, whilst another portion is combined with the oxygene of the atmospheric air contained in the water; and from the disengagement of the

* The water still being unity.

nitrogene of this air, arifes an increafed refiduum. *c. d.*

3. Probably in waters containing nitrogene,
hydrogene, and other gafes, abforbable only to
a flight extent, the apparent diminution will be
lefs, on account of the difengagement of thofe
gafes from the water, by the ftronger affinity of
nitrous gas for that fluid.

4. In water containing carbonic acid, and
probably fome other acid gafes, the diminution
will be fmall in proportion to the quantity of
gas contained in the water : the affinity of
this fluid for nitrous gas being diminifhed by its
greater affinity for the fubftance combined
with it. *e.*

The different diminution of nitrous gas when
agitated in different kinds of water, has been
long obferved by experimenters on the conftituent parts of the atmofphere, and various folutions
have been given of the phænomenon ; the moft
fingular is that of Humbolt.* He fuppofes

---

* He fays " 100 parties de gaz nitreux, (à o.14 d'azote) fe-
" couées avec l'eau diftillée, récemment cuite, diminuent en

that the apparent diminution of nitrous gas is lefs in fpring water than diftilled water; on account of the decompofition of the carbonate of lime contained in the fpring water, by the nitrous acid formed from the contact of nitrous gas with the water ; the carbonic acid difengaged from this decompofition increafing the refiduum.

This opinion may be confuted without even reference to my obfervations. It is, indeed,

" volume de 0.11, ou 0.12. Ce même gaz, en contact avec
" l'eau de puits, ne perd que 0.02. La caufe de cette dif-
" férence de 0.9, ou 0.10, ne doit pas être attribuée ni à
" l'impurité de l'air atmofphérique, contenu dans les inter-
" ftices de l'eau, ni à la décompofition de cette eau même.
" Elle n'eft qu'apparenté ; car l'acide nitrique, qui fe forme
" par le contact du gaz nitreux avec l'eau de puits, en dé-
" compofe le carbonate de chaux. Il fe dégage de l'acide
" carbonique, qui, en augmentant le volume du refidu, rend
" l'abforption du gaz nitreux moins fenfible. Pour déter-
" miner la quantité de cet acide carbonique, je lavai le
" réfidu avec de l'eau de chaux. Dans un grand nombre
" d'expériences, le volume diminua de 0.09, ou 0,07. Il
" faut en conclure que l'eau de puits abforbe réellement
" 9 + 2, ou 7 + 2 parties de gas nitreux, c'eft-à-dire, à
" peu-près la même quantité que l'eau diftillée."

Annales de Chimie, xxviii. pag. 154.

altogether unworthy of a philofopher, generally, acute and ingenious. He feems to have forgotten that carbonic acid is abforbable by water.

VII. *Of the abforption of Nitrous Gas, by folution of pale green Sulphate of Iron.*

a. The difcovery of the exact difference between the fulphates of iron, is owing to Prouft.* According to the ingenious refearches of this chemift, there exift two varieties of fulphate of iron, the green and the red. The oxide in the green fulphate contains $\frac{27}{100}$ oxygene. This falt, when pure, is infoluble in fpirit of wine; its folution in water is of a pale green color; it is not altered by the gallic acid, and affords a white precipitate with alkaline pruffiates.

The red fulphate of iron is foluble in alcohol and uncryftalizable; its oxide contains $\frac{48}{100}$ oxygene. It forms a black precipitate with the gallic acid, and with the alkaline pruffiates, a blue one.

* Nicholfon's Phil. Jour. No. 1, p. 453.

The common fulphates of iron generally confift of combinations of thefe two varieties in different proportions.

The green fulphate may be converted into the red by oxygenated muriatic acid or nitric acid. The common fulphate may be converted into green fulphate, by agitation in contact with fulphurated hydrogene.

The green fulphate has a ftrong affinity for oxygene, it attracts it from the atmofphere, from oxygenated marine acid, and nitric acid. The alkalies precipitate from it a pale green oxide, which if expofed to the atmofphere, rapidly becomes yellow red.

The red fulphate of iron has no affinity for oxygene, and when decompofed by the alkalies, gives a red precipitate, which undergoes no alteration when expofed to the atmofphere.*

b. The abforption of nitrous gas by a folution of fulphate of iron, was long ago difcovered by

---

* I have been able to make thefe obfervations on the fulphates of iron, moft of them after Prouft.

Prieftley: During this abforption, he remarked a change of color in the folution, analogous to that produced by the mixture of it with nitric acid.

This chemical fact has been lately applied by Humbolt, to the difcovery of the nitrogene generally mingled with nitrous gas.

Vauquelin and Humbolt have publifhed a memoir, on the caufes of the abforption* of nitrous gas by folution of fulphate of iron. They faturated an ounce and half of fulphate of iron in folution, with 180 cubic inches of nitrous gas.

Thus impregnated it ftrongly reddened tincture of turnfoyle; when mingled with fulphuric acid, gave nitric acid vapor; and faturated with potafh, ammoniacal vapor.

By analyfis, it produced as much ammoniac as that contained in 4 grains of ammoniacal muriate, and a quantity of nitric acid equal to that exifting in 17 grains of nitre. Hence they

* Annales de chimie, vol. xxviii. pag. 182.

concluded; that the nitrous gas and a portion of the water of the folution, had mutually decompofed each other; the oxygene of the water combining with the oxygene and a portion of the nitrogene of nitrous gas to form nitric acid; and its hydrogene uniting with the remaining nitrogene, to generate ammoniac.

They have taken no notice of the nature of the fulphate of iron employed, which was moft probably the common or mixed fulphate; nor of the attraction of the oxide of iron in this fubftance for oxygene.

c. Before I was acquainted with the obfervations of Prouft, the common facts relating to the oxygenation of vitriol of iron induced me to fuppofe, that the attraction of this fubftance for oxygene was in fome way connected with the procefs of abforption. The comparifon of the experiments of Humbolt and Vauquelin, with the obfervations of Prouft, enabled me to difcover the true nature of the procefs.

I procured a folution of red fulphate of iron, by paffing oxygenated muriatic acid through a folution of common fulphate of iron, till it gave only a red precipitate, when mingled with cauftic potafh. To nitrous gas confined by mercury, a fmall quantity of this folution, was introduced. On agitation, its color altered to muddy green; but the abforption that took place was extremely trifling: in half an hour it did not amount to ,2, the volume of the folution being unity, when it had nearly regained the yellow color.

I now obtained a folution of green ful-phate of iron, by diffolving iron filings in diluted fulphuric acid. The folution was agitated in contact with fulphurated hydrogene, and afterwards boiled; when it gave a white precipitate with pruffiate of potafh.

A fmall quantity of this folution agitated in nitrous gas, quickly became of an olive brown, and the gas was diminifhed with great rapidity; in two minutes, a quantity equal to four times the volume of the folution, had been abforbed.

Thefe facts convinced me that the folubility of nitrous gas in common fulphate of iron, chiefly depended upon the pale green sulphate contained by it; and that the attraction of one of the constituents of this substance, the green oxide of iron, for oxygene, was one of the causes of the phænomenon.

d. Green sulphate of iron rapidly decomposes nitric acid. It was consequently difficult to conceive how any affinities existing between nitrous gas, water, and green sulphate of iron, could produce the nitric acid found in the experiments of Vauquelin and Humbolt.

To ascertain if the presence of a great quantity of water destroyed the power of green sulphate of iron to decompose nitric acid, I introduced into a cubic inch of solution of green sulphate of iron, two drops of concentrated nitric acid.

The solution assumed a very light olive color; prussiate of potash mingled with a little of it, gave a dark green precipitate. Hence the nitric acid had been evidently decomposed. As no nitrous gas was given out, which is always the case when nitric acid is poured on cryſtaliſed, ſulphate of iron, I ſuſpected that a compleat decompoſition of the acid had taken place; but when the ſolution was heated, a few minute globules of gas were liberated, and it gradually became ſlightly clouded.

Having often remarked that no precipitation is ever produced during the converſion of green ſulphate of iron into red, by oxygenated muriatic acid, or concentrated nitric acid, I could refer the cloudineſs to no other cauſe than to the formation of ammoniac.

To aſcertain if this ſubſtance had been produced, a quantity of ſlacked cauſtic lime was thrown into the ſolution. On the application of heat, the ammoniacal ſmell was diſtinctly perceptible, and the vapor held over orange nitrous acid, gave denſe white fumes.

e. When I conſidered this fact of the decompoſition of nitric acid and water by the ſolution of green ſulphate of iron, and the change of color effected in it by the abſorption of nitrous gas, exactly analogous to that produced by the decompoſition of nitric acid; I was induced to believe that the nitric acid found in the analyfis of Vauquelih and Humbolt, had been formed by the combination of fome of the nitrous gas thrown into the folution with the oxygene of the atmofphere: and that the abforbability of nitrous gas, by folution of green fulphate of iron, was owing to a decompofition produced by the combination of its oxygene with the green oxide of iron, and of its nitrogene with the hydrogene difengaged from water, decompounded at the fame time.

To afcertain this, I procured a quantity of nitrous gas: it was fuffered to remain in contact with water for fome hours after its production. Transferred to the mercurial apparatus, it gave no white vapor when placed in contact with folution of ammoniac; and confequently held no nitric acid in folution.

Into a graduated jar filled with mercury, a cubic inch of concentrated folution of pure green fulphate of iron was introduced, and 7 cubic inches of nitrous gas admitted to it. The folution immediately became dark olive at

the edges, and on agitation this color was diffufed through it. In 3 minutes, when near 5¾
cubic inches had been absorbed; the diminution
ceafed. The folution was now of a bright olive
brown, and tranfparent at the edges. After it had
refted for a quarter of an hour, no farther abforption was obferved ; the color was the
fame, and no precipitation could be perceived.
A little of it was thrown into a fmall glafs tube,
under the mercury, and examined in the atmofphere. Its tafte was rather more aftringent
than that of folution of green fulphate ; it
did not at all alter the color of red cabbage
juice. When a little of it was poured on the
mercury, it foon loft its color, its tafte became
acid, and it quickly reddened cabbage juice,
even rendered green by an alkali.

To the folution remaining in the mercurial
jar, a fmall quantity of pruffiate of potafh was
introduced, to afcertain if any red fulphate of
iron had been formed ; but inftead of the production of either a blue, or a white precipitate,
the whole of the folution became opaque, and
chocolate colored.

Surprifed at this appearance, I was at firft induced to fuppofe, that the ammoniac formed by the nitrogene of the nitrous gas and the hydrogene of the water, had been fufficient to precipitate from the fulphuric acid, the red oxide of iron produced, and that the color of the mixture was owing to this precipitation. To diffolve any uncombined oxide that might exift in the folution, I added a very minute quantity of diluted fulphuric acid; but little alteration of color was produced. Hence, evidently, no red oxide had been formed.

This unexpected refult obliged me to theorife a fecond time, by fuppofing that nitrate of ammoniac had been produced, which by combining with the white pruffiate of iron, generated a new combination. But on mingling together green fulphate of iron, pruffiate of potafh, and nitrate of ammoniac in the atmofphere, the mixture remained perfectly white.

To afcertain if any nitric acid exifted, combined with any of the bafes, in the impregnated folution, I introduced into it an equal bulk of diluted sulphuric acid : it became rather paler ; but no green or blue tinge was produced.

That the pruffic acid had not been decompofed, was evident from the bright green produced, when less than a grain of dilute nitric acid was admitted into the folution.

f. From thefe experiments it was evident, that no red fulphate of iron, or nitric acid, and confe-quently no ammoniac, had been produced after the abforption of nitrous gas by green fulphate of iron. And when I compared them with the obfervations of Prieftley, who had expelled by heat a minute quantity of nitrous gas from an impregnated folution of common fulphate of iron, and who found common air phlogifticated by ftanding in contact with it, I began to fuf-pect that nitrous gas was fimply diffolved in the folution, without undergoing decompofi-tion.

g. To determine more accurately the nature of the procefs, I introduced into a mercurial cylinder 410 grains of folution of green fulphate of iron, occupying a fpace nearly equal to a cubic inch and quarter ; it was saturated with nitrous gas, by abforbing 8 cubic inches. This faturated folution exhibited the fame appearance as the laft ; and after remaining near an hour untouched, had evidently depofited no oxide of iron, nor gained any acid properties.

Into a fmall mattrafs filled with mercury, having a tight ftopper with a curved tube adapted to it, the greater part of this folution was introduced ; judging from the capacity of the mattrafs, about 50 grains of it might have been loft. To prevent common air from coming in contact with the folution, the ftopper was introduced into the mattrafs under the mercury ; the curved tube connected with a graduated cylinder filled with that fubftance ; and the mattrafs brought over the fide of the mercurial trough. But in fpite of thefe precautions a large globule of common air got into the top of the mattrafs, from the curvature of the tube. When the heat of a fpirit lamp was applied to the folution, it gave out gas with great rapidity, and gradually loft its color. When 5 cubic inches were collected it became, perfectly pale green, whilst a yellow red precipitate was deposited, on the bottom of the mattrass.

On-pouring a little of the clear solution, into prussiate of potash, it gave only white prussiate of iron.

But on introducing a particle of sulphuric acid into the solution, sufficient to dissolve some of the red precipitate, and then pouring a little of it into a solution of prussiate of potash, it gave a fine blue prussiate of iron.

Hence the red precipitate was evidently red yellow oxide of iron.

I now examined the gas, suspecting that it was nitrous oxide. On mingling a little of it with atmospheric air, it gave red vapor, and diminished. Solution of sulphate of iron introduced to the remainder, almost wholly absorbed it : the small residual globule of nitrogene could not equal one thirtieth of a cubic inch.

Consequently it was nitrous gas, nearly pure.

Caustic potash was now introduced into the solution, till all the oxide of iron was precipitated. The solution, when heated, gave a

ſtrong ſmell of ammoniac, and denſe white
fumes when held over muriatic acid. It was
kept at the heat of ebullition till the evaporation had been nearly compleated. Sulphuric
acid poured upon the reſiduum gave no yellow
fumes, or nitric acid vapor in any way perceptible; even when heated and made to boil,
there was no indication of the production of
any vapor, except that of the ſulphuric acid.

b. This experiment, compared with the others,
ſeemed almoſt to prove, that nitrous gas
combined with ſolution of pale green ſulphate
of iron, at the common temperature, without
decompoſition; and that when the impregnated
ſolution was heated, the greater portion of gas
was diſengaged, whilſt the remainder was decompounded by the green oxide of iron; which
attracted at the ſame time oxygene from the
water and the nitrous gas; whilſt their other
conſtituent principles, hydrogene and nitrogene,
entered into union as ammoniac.

Whilſt, however, I was reaſoning upon this
ſingular chemical change, as affording pre-

fumptive proofs in favor of the exertion of fimple affinities by the conftituent parts of compound fubftances, a doubt concerning the decompofition of the nitrous gas occurred to me. As near as I could guefs at the quantity of nitrous gas contained by the impregnated folution, at leaft ¾ of it muft have been expelled undecompounded.

More than a quarter of a cubic inch of common air had been prefent in the mattrafs: the oxygene of this common air muft have combined with the nitrous gas, to form nitric acid. Might not this nitric acid have been decompofed, and furnifhed oxygene to the red oxide of iron, and nitrogene to the fmall quantity of ammoniac found in the folution, as in *d?*

i. I now introduced to a folution of green fulphate confined by mercury, nitrous gas, perfectly free from nitric acid. When the folution was faturated, a portion of it was introduced into a fmall mattrafs filled with dry mercury, in the mercurial trough. The curved tube was clofed by a fmall cork at the top, and filled

with nitrous gas; it was then adapted to the
mattrafs, which was raifed from the trough, and
the folution thus effectually preferved from the
contact of the atmofphere.

When the heat of a fpirit. lamp was applied
to the mattrafs, it began to give out gas with
great rapidity. After fome time the folution
loft its dark color, and became turbid. When
the production of nitrous gas had ceafed, it was
fuffered to cool. A copious red precipitate
had fallen down; which, examined by the fame
tefts as in the laft experiment, proved to be
red oxide of iron.

The folution treated with lime, as before,
gave ammoniac; but with fulphuric acid, not
the flighteft indications of nitric acid.

k. Having thus procured full evidence of the
decompofition of nitrous gas in the heated folution, in order to gain a more accurate acquaintance with the affinities exerted, I endeavoured to afcertain the quantity of nitrous gas
decompofed by a given folution, under known
circumftances.

Into a cylinder of the capacity of 20 cubic inches, inverted in mercury, 1150 grains of folution of green fulphate of iron, of fpecific gravity 1.4, were introduced. Nitrous gas was admitted to it, and after fome time 21 cubic inches were abforbed.

The impregnated folution was thrown into a mattrafs, in the fame manner as in the laft experiment, and the fame precautions taken to preferve it from the contact of atmofpheric air. A quantity was loft during the procefs of tranfferring, which, reafoning from the fpace occupied in the mattrafs by the remaining portion, as determined by experiment afterwards, muft have amounted nearly to 240 grains.

The curved tube from the mattrafs was now made to communicate with the mercurial airholder. By the application of heat 12,5 cubic inches of nitrous gas were collected, after the common temperature had been reftored to the mattrafs; which was fuffered to remain in communication with the conducting tube.

The folution was now pale green, that is, of its

natural color, and a confiderable quantity of red oxide of iron had been depofited.

Solid cauftic potafh was introduced into it, till all the green oxide of iron had been precipi- tated, and till the folution rendered green, red cabbage juice.

A tube was now accurately connected with the mattrafs, bent, and introduced into a fmall quantity of diluted fulphuric acid. Nearly half of the fluid in it was flowly diftilled into the fulphuric acid, by the heat of a fpirit lamp. The impregnated acid evaporated at a heat above 212°, and gave a fmall quantity of cryf-. talifed falt, which barely amounted to two grains and quarter ; it had every property of fulphate of ammoniac. Sulphuric acid in excefs was poured on the refiduum, and the whole diftilled by a heat not exceeding 300°, into a fmall quan- tity of water. This water, after the procefs, tafted ftrongly of fulphuric acid ; it had no peculiar odor. Tin-thrown into it when heated, was not perceptibly oxydated ; mingled with ftrontitic lime ,water, it gave a copious white precipitate, and after the precipitation became almoſt taſteleſs. Hence it evidently contained no nitric acid.

The 12,5 cubic inches of undecompounded gas that came over were examined; and accounting for the ſmall quantity of common air previouſly contained in the airholder, muſt have been almoſt pure.

1. Now ſuppoſing 927 grains of the impregnated ſolution (including the weight of the nitrous gas), to have been operated upon, this muſt have contained about 16,7 cubic inches of nitrous gas. But 12,5 cubic inches eſcaped undecompounded : hence 4,2 were decompoſed; and theſe weigh 1,44 grains, and are compoſed of ,8 oxygene, and ,64 nitrogene.*

Conſequently, the nitrous gas muſt have furniſhed ,8 of oxygene to the green oxide of iron.

But ,64 of nitrogene require ,15 of hydrogene to form ,79 of ammoniac :† conſequently 1 of

---

* Diviſion IV. Section 5.

† Diviſion II. Section 1.

water was decompounded, and this furnished
,85 of oxygene to the green oxide of iron.

The green oxide of iron contains $\frac{27}{100}$ oxygene;
the red $\frac{48}{100}$. But the whole quantity of oxygene
fupplied from the water and nitrous gas is
8 + 85 = 1,65 ; and calculating on the difference of the compofition of the red and green
oxide of iron, 5,7 grains of red oxide muft have
been depofited, and confequently thefe would
faturate as much acid as ,79 grains of ammoniac,
or 4,1 grains of green oxide of iron.*

And fuppofing the ammoniac in fulphate of
ammoniac to be to the acid as 1 is to 3,† 3.2
grains of fulphate of ammoniac muft have been
formed, containing about 2,4 grains acid ; and
then 6,5 grains of green fulphate of iron muft
have been decompofed.

Hence we gain the following equation :

---

* No precipitation takes place during the converfion of
folution of green fulphate into red ; and the acid appears
faturated.

† Divifion II, Section 6.

6,5 green f. = 2,41 ful. acid + 4,1 gr. ox. iron.
+
1,44 nit. gas = ,64 nitrogene + ,8 oxygene.
+
1 water = ,85 oxygene + ,15 hydrogene.

equal

3,2 ful. am. = 2,41 f. acid + ,64 nit. + ,15 hyd.
+
5,7 r. ox. iron = 4,1 gr. ox. iron + 1,6 oxyg.

Though the estimation of the quantities in this equation must not be considered as strictly accurate, on account of the degree of uncertainty that remains concerning the exact numerical expression of the quantities of the constituents of water, ammoniac, and the other compound bodies employed; yet as founded on a simple quantity, that is, the nitrous gas decomposed, it cannot be very distant from the truth.

The sulphate of ammoniac given by experiment, is considerably less than that which was really produced; much of it was probably carried off during the evaporation of the superabundant acid.

The conclufions that may be drawn from this experiment, afford a ftriking inftance of the importance of the application of the fcience of quantity to the chemical changes ; for the data, being one chemical fact, the decompofition of a given quantity of nitrous gas by known agents ; the compofition of nitrous gas, of water, ammoniac; the oxides of iron, and fulphate of ammoniac ; we are able not only to determine the quantities of the fimple conftituents that have entered into new arrangements, but likewife the compofition of two compound bodies, the green and red fulphates of iron.*

m. Though from the experiments in e it appeared that no decompofition of nitrous gas had been produced during, or even after its abforption, by folution of fulphate of iron at the common temperature ; yet a fufpicion that it might take place flowly, and that

* According to the eftimation in the equation, 6.5 of dry green fulphate of iron contain 4.1 green oxide of iron, and 2.4 of Kirwan's real fulphuric acid ; and 8.1 red fulphate of iron, contain 2.4 acid, and 5.7 red oxide of iron.

indications of it might be given by depofition, induced me to examine minutely two
impregnated folutions, one of which had been
at reft, confined by mercury, for 19 hours, and
the other for 27. In neither of them could I
difcover any depofition, or alteration of color,
which might denote a change.

Two cubic inches of oxygene were admitted
to half a cubic inch of one of thefe folutions.
The oxygene was flowly abforbed, and the folution gradually loft its color.

To afcertain if during the converfion of the
nitrous gas held in folution by fulphate of
iron, into nitric acid, by the oxygene of the
atmofphere at the common temperature, any
water was decompofed ; I fuffered an impregnated folution, weighing nearly two ounces, to
remain in contact with the atmofphere at 57°—
62°, till it was become perfectly pale. It then
had a ftrong acid tafte, effervefced with carbonate of potafh, and gave a blue precipitate
with pruffiate of potafh.—It was faturated
with quicklime, and heated : flight indications

of the prefence of ammoniac were perceived.

As in this experiment the nitric acid had been moft probably decompofed by the green oxide of iron, as in *f,* I fent oxygenated muriatic acid through an impregnated folution, till all the green oxide of iron was converted into red, and all the nitrous gas into nitric acid.

This folution faturated with potafh, and heated, gave no ammoniacal fmell.

From thefe experiments we may conclude,

1ft. That folution of red fulphate of iron has little or no affinity for nitrous gas*; and that folution of common fulphate abforbs nitrous gas only in proportion as it contains green fulphate.

2dly. That folutions of green fulphate of iron diffolve nitrous gas in quantities proportionable to their concentration, without effecting

---

* The muddy green color produced in a folution of red fulphate of iron agitated in nitrous gas, depended upon impurities in the mercury. I have fince found, that when the folution is completely oxygenated, the diminution is barely perceptible.

any decompofition of it at common temperatures.
And the folubility of nitrous gas in folution
of green fulphate, may be fuppofed to depend
on an equilibrium of affinity, produced by the
following fimple attractions :

1. That of green oxide of iron for the oxygene of nitrous gas and water.

2. That of the hydrogene of the water
for the nitrogene of the nitrous gas.

3. That of the principles of the fulphuric acid, for nitrogene and hydrogene.

3dly. That at high temperatures, that is,
from 200° to 300°, the equilibrium of affinity
producing the binary combination between
nitrous gas and folution of green fulphate of
iron is deftroyed ; the attraction of the green
oxide of iron for oxygene beingincreafed ; whilft
probably that of nitrogene for hydrogene is
diminifhed.

Hence the nitrous gas is either liberated,* in

---

* Perhaps the liberation of nitrous gas from the folution

confequence of the affinity between oxygene
and hydrogene, and oxygene and nitrogene not
following the fame ratio of alteration on increafed temperature; or decompofed, becaufe
at a certain temperature the green oxide exerts
fuch affinities upon water and nitrous gas, as to
attract oxygene from both of them to form red
oxide; whilft the ftill exifting affinity between
the hydrogene of the one, and the nitrogene of
the other, difpofes them to combine to form
ammoniac.

4thly. That the change of color produced by
introducing nitric acid to folution of common
fulphate of iron, exactly analogous to that occafioned in it by impregnation with nitrous gas,
is owing to the decompofition of the acid, by
the combination of its oxygene with the green

***

takes place at a lower temperature than its dccompofition.
I have always obferved that the quantity of yellow precipitate is greater when the folution is rapidly made to
boil. Were it poffible to heat it to a certain temperature at once, probably a compleat decompofition
would take place.

oxide of iron, and of its nitrous gas with the solution.

5thly. That nitrous gas in combination with solution of green sulphate of iron, is capable of exerting a strong affinity upon free or loosely combined oxygene, and of uniting with it to form nitric acid.

.n. The products obtained from a solution of sulphate of iron saturated with nitrous gas, by Vauquelin and Humbolt, and their consequent mistake with regard to the nature of the procefs of abforption,* must have arisen from expofure of their impregnated folution to the atmofphere.

Indeed, from the acidity of it, on examination, from the fmall portion of ammoniac, and the large quantity of nitric acid obtained, it appears moft probable that the whole of the nitrous gas employed was converted into nitric acid, by combining with atmofpheric oxygene; for no nitric acid could have been obtained in

* Annales de Chimie. T. 38, pag. 187.

the mode in which they operated, unlefs the green oxide of iron in the folution had been previoufly converted into red.

VIII. *On the abforption of Nitrous Gas by folution of green Muriate of Iron.*

a. The analogy between the affinities of the conftituents of the muriate and fulphate of iron, induced me to conjecture that they poffeffed fimilar powers of abforbing nitrous gas; and I foon found that this was actually the cafe; for on agitating half a cubic inch of folution of muriated iron, procured by diffolving iron filings in muriatic acid, in nitrous gas, the gas was abforbed with great rapidity, whilft the folution affumed a deep and bright brown tinge.

b. Prouft,* who as I have before mentioned, fuppofes the exiftence of two oxides of iron only,

---

* Annales de Chimie, xxiii. pag. 85; or Nicholfon's Phil. Journal vol. i. pag. 45.

one containing $\frac{27}{100}$ oxygene, the other $\frac{48}{100}$, has assumed, that the muriatic acid, and most other acids as well as the sulphuric, are capable of combining with these oxides, and of forming with each of them a distinct salt. He has, however, detailed no experiments on the muriates of iron.

As these salts are still more distinct from each other in their properties than the sulphates, and as these properties are connected with the phænomenon of the absorption and decompoſition of nitrous gas, I shall detail the obfervations I have been able to make upon them.

c. When iron filings have been diffolved in pure muriatic acid, and the folution preferved from the contact of air, it is of a pale green color, and gives a white precipitate with alkaline pruffiates. The alkalies throw down from it a light green oxide of iron.

When evaporated, it gives cryftals almoft white, which are extremely foluble in water; but infoluble in alcohol.

The folution of green muriate of iron has a

great affinity for oxygene, and attracts it from the atmofphere, from nitric acid, and probably from oxygenated muriatic acid.

When red oxide of iron is diffolved in muriatic acid, or when nitric acid is decompofed by folution of green muriate of iron ; the red muriate of iron is produced. The folution of this falt is of a deep brown red, its odor is peculiar; and its tafte, even in a very diluted ftate, highly aftringent. It acts upon animal and vegetable matters in a manner fomewhat analogous to the oxygenated muriatic acid, rendering them yellowifh white, or yellow.*

Sulphuric acid poured upon it, produces a fmell refembling that of oxygenated muriatic acid. Evaporated at a low temperature, it gives an uncryftalifable dark orange colored falt, which is foluble in alcohol, and when decompofed by the alkalies, gives a red precipitate. With pruffiate of potafh it gives pruffian blue.

---

* Probably by giving them oxygene ; whereas the green muriate and fulphate blacken animal fubftances ; moft likely by abftracting from them oxygene.

The common muriate of iron confifts of different proportions of thefe two falts. It may be converted into red muriate by concentrated nitric acid, or into green by fulphurated hydrogene.

d. To afcertain if folution of red muriate of iron was capable of abforbing nitrous gas, I introduced into a jar filled with mercury, a cubic inch of nitrous gas, and admitted to it nearly half a cubic inch of folution of red muriate of iron. No difcoloration took place. By much agitation, however, an abforption of nearly 2 was produced, and the folution became of a muddy green. But this change of color, and probably the abforption, was in confequence of the oxydation of either the mercury, or fome imperfect metals combined with it, by the oxygene of the red muriate. For I afterwards found, that precifely the fame change of color was produced when a folution was agitated over mercury.

e I introduced to a cubic inch of concentrated folution of green muriate of iron, 7 cubic

inches of nitrous gas, free from nitric acid ; the
folution inftantly became colored at the edges,
and, on agitation abforbed the gas with much
greater rapidity than even fulphate of iron ; in
a minute, only a quarter of a cubic inch remained.

The folution appeared of a very dark brown,
but, evidently no precipitation had taken place
in it, and the edges, when viewed againft the
light, were tranfparent and puce colored.

Five cubic inches more of nitrous gas were
now diffolved in the folution. The intenfity
of the color increafed, and after an hour no
depofition had taken place. A little of it was
then examined in the atmofphere ; it had a
much more aftringent tafte than the unimpregnated folution, and effected no change in red
cabbage Juice. When pruffiate of potafh was
introduced into it, its color changed to olive
brown. A few drops of the folution, that had
accidentally fallen on the mercury, foon became
colorlefs, and then effervefced with carbonate
of potafh, and tafted ftrongly acid.

The remainder of the impregnated solution, which muft have nearly equalled ;75 cubic inches, was introduced into a mattrafs, having a ftopper and curved tube, as in the experiments on the folution of fulphate of iron; great care being taken to preferve it from the contact of air.

The mattrafs was heated by a fpirit lamp, the curved tube being in communication with a mercurial cylinder. Near 8 cubic inches of nitrous gas were collected; when the folution became of a muddy yellow. It was fuffered to cool, and examined. A fmall quantity of yellow precipitate covered the bottom of the mattrafs; the fluid was pellucid, and light green. A little of it thrown on pruffiate of potafh, gave a white precipitate, colored by ftreaks of light blue. When the yellow precipitate was partly diffolved by fulphuric acid, a drop of the folution, mingled with pruffiate of potafh, gave a deep blue green.

Hence, evidently, the precipitate was red oxide of iron.

Cauftic potafh in excefs was introduced into the remainder of the folution, and it was heated. It gave an evident fmell of ammoniac, and denfe white fumes, when held over ftrong phlo-gifticated nitrous acid.

When half of it was evaporated, fulphuric acid in excefs was poured on the remainder ; muriatic acid was liberated, not perceptibly combined with any nitric acid.

f. In an experiment that I made to afcertain the quantity of nitrous gas capable of combining with folution of green muriate of iron ; I found that ,75 cubic inches of faturated folution ab-forbed about 18 of nitrous gas, which is nearly double the quantity combinable with an equal portion of the ftrongeft folution of fulphate of iron. A part of this impregnated folution, heated flowly, gave out more gas in proportion to the quantity it contained, than the laft, and confequently produced lefs precipitate ; fo that I am inclined to fuppofe it probable, that at a certain temperature, all the diffolved nitrous gas may be difpelled from a folution.

From thefe experiments we may conclude,

1ft. That the folution of green muriate of iron abforbs nitrous gas in confequence of nearly the fame affinities as folution of green fulphate of iron ; its capability of abforbing larger quantities depending moft probably on its greater concentration (that is, on the greater folubility of the muriate of iron), and perhaps, in fome meafure, on a new combining affinity, that of muriatic acid for oxygene.

2dly. That at certain temperatures nitrous gas is either liberated from folution of green muriate, or decompofed, by the combination of its oxygene with green oxide of iron, and of its nitrogene with hydrogene, produced by water decompounded by the oxide at the fame time.

IX. *Abforption of Nitrous Gas by Solution of Nitrate of Iron.*

a. As well as two fulphates and two muriates

of iron, there exift two nitrates.* When concentrated nitric acid, is made to act upon iron,
nitrous gas is difengaged with great rapidity,
and with great increafe of temperature : the
folution affumes a yellowifh tinge, and as the
procefs goes on, a yellow red oxide is precipitated.

Nitrate of iron made in this way, gives a
bright blue mingled with pruffiate of potafh,
and decompofed by the alkalies, a red precipitate. Its folution has little or no affinity for
nitrous gas.

. b. When very dilute nitric acid, that is,
fuch as of fpecific gravity 1,16, is made to
oxydate iron, without the affiftance of heat,
the folution gives out no gas for fome time, and
becomes dark olive brown : when neutralifed
it gives, decompofed by the alkalies, a light green
precipitate ; and mingled with pruffiate of potafh, pale green pruffiate of iron.

---

* The exiftence of green nitrate was not fufpected by
Prouft.

It owes its color to the nitrous gas it holds in solution. By expofure to the atmofphere it becomes pale, the nitrous gas combined with it being converted into nitric acid.

It is then capable of abforbing nitrous gas, and confifts of pale nitrate of iron, mingled with red nitrate.

I have not yet obtained a nitrate of iron giving only a white precipitate with pruffiate of potafh, that is, fuch as contains *only* oxide of iron at its minimum of oxydation; for when pure green oxide of iron is diffolved by very dilute nitric acid, a fmall quantity of the acid is generally decompofed, which is likewife the cafe in the decompofition of nitre by green fulphate of iron. The folutions of nitrate of iron, however, procured in both of thefe modes, abforb nitrous gas with rapidity, and by fulphurated hydrogene might probably be converted into pale nitrate.

As it is impoffible to obtain concentrated folutions of pale nitrate of iron, chiefly containing green oxide, its powers of abforbing nitrous gas cannot be compared with the muriatic and sulphuric solutions, unless they are made of nearly the same specific gravity.

Nitrous gas is disengaged by heat from the impregnated solution of nitrate of iron, at the same time that much red oxide of iron is precipitated. Whether any nitrous gas is decompofed, I have not yet ascertained; for when unimpregnated pale nitrate of iron is heated, a part of the acid, and of the water of the solution, is decompofed by the green oxide of iron;* and in consequence ammoniac, and red nitrate of iron formed, whilst red oxide is precipitated.

X. *Absorption of Nitrous Gas by other Metallic Solutions.*

a. White pruffiate of iron in contact with water absorbs nitrous gas to a great extent, and

---

* In this process nitrous oxide is sometimes given out, as will be seen hereafter.

becomes dark chocolate.*

b. Concentrated folution of fulphate of tin, *probably* at its minimum of oxydation, abforbs one eighth of its bulk of nitrous gas, and becomes brown, without depofition.

c. Solution of fulphate of zinc‡ abforbs about one tenth of its volume of nitrous gas, .and becomes green.

d. Solution of muriate of zinc‡ abforbs nearly the fame quantity, and becomes orange brown.

e. Thefe are all the metallic fubftances on which I have experimented. It is more than probable that there exift others poffeffing fimilar powers of abforbing nitrous gas.

Whenever the metals capable of decompofing water exift in folutions at their minimum of oxydation, the affinities exerted by them on

***

* Hence we learn why no nitrous gas is difengaged when impregnated folution of fulphate of iron is decompofed by pruffiate of potafh, as in Div. IV. Sec. vii.

‡ In both of thefe folutions the metal is at its minimum of oxydation. The abforption of a fmall quantity of nitrous gas by white vitriol was obferved by Prieftley.

nitrous gas and water, will be such as to produce combination. The powers of metallic solutions to combine with nitrous gas at common temperatures, as well as to decompose it at higher temperatures, will probably be in the ratio of the affinity of the metallic oxides they contain, for oxygene.

XI. *The action of Sulphurated Hydrogene on solution of Green Sulphate of Iron, impregnated with Nitrous Gas.*

a. In an experiment on the absorption of nitrous gas by solution of green sulphate of iron, I introduced an unboiled solution of common sulphate, deprived of red oxide of iron by sulphurated hydrogene, into a jar filled with nitrous gas; the absorption took place as usual, and nearly six of gas entered into combination, the volume of the solution being unity. On applying heat to a part of this impregnated solution, the whole of the nitrous gas it contained (as nearly as I could guess), was expelled undecompounded, and no yellow precipitate produced. Pruſſiate of potaſh poured into it gave only white pruſſiate of iron ; and when it was, heated with lime, no ammoniacal ſmell was perceptible.

I could refer this phænomenon to no other, cauſe than to the exiſtence of a ſmall quantity of ſulphurated hydrogene in the ſolution. That this was the real cauſe I found from the following experiment.

b. One part of a ſolution of green ſulphate of iron, formed by the agitation of common ſulphate of iron in contact with ſulphurated hydrogene, was boiled for ſome minutes to expel the ſmall quantity of gas retained, by it undecompounded. It had, then no peculiar ſmell, and gave a white pruſſiate of iron with pruſſiate of potaſh ; the other part had a faint odor of ſulphurated hydrogene, and gave a dirty white precipitate with pruſſiate of potaſh; Nearly equal quantities of each were ſaturated, with nitrous gas, and heated. The unboiled impregnated ſolution gave out all its nitrous gas undecompounded ; whilſt in the boiled ſolution it was partly decompoſed, yellow precipitate and ammoniac being formed.

c. This ſingular phænomenon of the power of a minute quantity of ſulphurated hydrogene, in preventing the decompoſition of nitrous gas and water, by green oxide of iron, will moſt probably take place in other impregnated ſolutions. It ſeems to depend on the ſtrong affinity of the hydrogene of ſulphurated hydrogene for oxygene.

XII. *Additional Obſervations.*

a. For ſeparating nitrous gas from gaſes abſorbable to no great extent by water; a well boiled ſolution of green muriate of iron ſhould be employed. Nitrous gas agitated in this is rapidly abſorbed, and it has no affinity for, or action on, nitrogene, hydrogene, or hydrocarbonate.

b. Nitrous gas carefully obtained from mercury and nitric acid, when received under mer-

N

cury, or boiled water, and absorbed by solution of green muriate, or sulphate of iron, rarely leaves a residuum of $\frac{1}{200}$ of its volume: preserved over common water, and absorbed, the remainder is generally from $\frac{1}{4}$ to $\frac{1}{90}$, from the nitrogene disengaged by the decomposition of the common air contained in the water.

c. The nitrous gas carefully obtained from the decomposition of nitric acid of 1.26, by copper, I have hardly ever found to contain more than from $\frac{1}{30}$ to $\frac{1}{5}$ nitrogene, when received through common water: when boiled water is employed, the residuum is nearly the same as that of nitrous gas obtained from mercury.

d. Consequently the gas from those two solutions may be used in common. It is more than probable, that the small quantities of nitrogene generally mingled with nitrous gas from copper and mercury, arise either from the common air of the vessels in which it was produced, or that of the water over which it was received. There is no reason for supposing that it is generated by a complete decomposition of a portion of the acid.*

e. Whenever nitrous oxide is mingled with nitrous gas and nitrogene, it muſt be ſeparated by well boiled water ; and after the corrections are made for the quantity of air diſengaged from the water, the nitrous gas abſorbed by the muriatic ſolution.

---

* Humbolt, who is the firſt philoſopher that has applied the ſolution of ſulphate of iron to aſcertain the purity of nitrous gas, aſſerts that he uniformly found nitrous gas obtained from ſolution of copper in nitrous acid, to contain from ſix tenths to one tenth nitrogene.

Annales de Chimie, vol. xxviii. pag. 147.

DIVISION V.

EXPERIMENTS and OBSERVATIONS on the production of NITROUS OXIDE from NITROUS GAS and NITRIC ACID, in different modes.

I. Preliminaries.

a. The opinions of Prieftley[*] and Kirwan,[†] relating to the caufes of the converfion of nitrous gas into nitrous oxide, were founded on the theory of phlogifton. The firft of thefe philofophers obtained nitrous oxide by placing nitrous gas in contact with moiftened iron filings, or the alkaline fulphures. The laft by expofing it to fulphurated hydrogene.

The Dutch chemifts,[‡] the lateft experi-

---

[*] Vol. ii. pag. 55. [†] Phil. Tranf. vol. lxxvi. pag. 133.

[‡] Journal de Phyfique, tom. xliii. 323.

mentalifts on nitrous oxide, have fuppófed that
the production of this fubftance depends upon
the fimple abftraction of a portion of the oxygene of nitrous gas. They obtained nitrous
oxide by expofing nitrous gas to muriate of
tin, to copper in folution of ammoniac, and
likewife by paffing it over heated fulphur.

The diminution of volume fuftained by
nitrous gas during its converfion into nitrous
oxide, has never been accurately afcertained;
it has generally been fuppofed to be from two
thirds to eight tenths.

b. Nitrous gas may be converted into nitrous
oxide in two modes.

Firft, by the fimple abftraction of a portion
of its oxygene, by bodies poffeffing a ftrong
affinity for that principle, fuch as alkaline
fulphites, muriate of tin, and dry fulphures.

Second, by the combination of a body
with a portion both of its oxygene and
nitrogene, fuch as hydrogene, when either in a
nafcent form, or a peculiar ftate of combination.

c. Each of thefe modes will be diftinctly treated of; and to prevent unneceffary repetitions, I fhall give an account of the general manner in which the following experiments on the converfion of nitrous gas into nitrous oxide, have been conducted.

Nitrous gas, the purity of which has been accurately afcertained by folution of muriate of iron, is introduced into a graduated jar filled with dry mercury. If a fluid fubftance is defigned for the converfion of the gas into nitrous oxide, it is heated, to expel any loofely combined air which might be liberated during the procefs; and then carefully introduced into the jar, by means of a fmall phial. After the procefs is finifhed, and the diminution accurately noted, the nitrous oxide formed is abforbed by pure water. If any nitrous gas remains, it is condenfed by folution of muriate of iron; other refidual gafes are examined by the common tefts. The quantity of nitrous oxide diffolved by the fluid is determined by a comparative experiment; and the corrections for tempera-

ture and preffure being gueffed at, the conclufions drawn.

If a folid fubftance is ufed, rather more nitrous
gas than that defigned for the converfion, is
introduced into the jar. The fubftance is
brought in contact with the gas, by being
carried under the mercury; and as a little common air generally adheres to it, a fmall portion
of the nitrous gas is transferred into a graduated
tube, after the infertion, and its purity afcertained. In other refpects the procefs is conducted as mentioned aboye.

II. *Of the converfion of Nitrous gas into Nitrous Oxide, by Alkaline Sulphites.*

The alkaline fulphites, particularly the fulphite of potafh, convert nitrous gas into nitrous
oxide, with much greater rapidity than any
other bodies.

At temperature 46°, 16 cubic inches of
nitrous gas were converted, in lefs than an
hour, into 7,8 of nitrous oxide, by about 100

grains of pulverifed fulphite of potafh, contain-ing its water of cryftalifation. No fenfible increafe of temperature was produced during the procefs, no water was decompofed, and the quantity of nitrogene remaining after the experiment, was exactly equal to that previoufly contained in the nitrous gas.

The nitrous oxide produced from nitrous gas by fulphite of potafh, has all the properties of that generated from the decompofition of nitrate of ammoniac. It gives, as will be feen hereafter, the fame products by analyfis. Phofphorus, the taper, fulphur, and charcoal, burn in it with vivid light. It is abforbable by water, and capable of expulfion from it unaltered, by heat.

Nitrous gas is converted into nitrous oxide by the alkaline fulphites with the fame readinefs, whether expofed to the light, or deprived of its influence.

The folid fulphites act upon nitrous gas much more readily than their concentrated folutions; they fhould however always be fuffered to retain their water of cryftalifation, or otherwife they attract moifture from the gas, and render it drier, and in confequence more condenfed than it would otherwife be. In cafe per- fectly dry fulphites are employed, the gas fhould be always faturated with moifture after the experiment, by introducing into the cylin- der a drop of water.

The fulphites, after expofure to nitrous gas, are either found wholly, or partially, converted into fulphates. Confequently the converfion of nitrous gas into nitrous oxide by thefe bodies, fimply depends on the abftraction of a portion of its oxygene; the nitrogene and remaining oxygene affuming a more condenfed ftate of exiftence.

If we reafon from the different fpecific gra- vities of nitrous oxide and nitrous gas, as com- pared with the diminution of volume of nitrous gas, during its converfion into nitrous oxide, 100 parts of nitrous gas, fuppcfing the former eftimation of the compofition of nitrous oxide given in Divifion III, accurate, would confift of 54 oxygene, and 46 nitrogene; which is not far from the true eftimation. Or affuming the compofition of nitrous gas, as given in Divifion IV, it would appear from the diminution, that 100 parts of nitrous oxide confifted of 38 oxygene, and 62 nitrogene.

III. *Converfion of Nitrous Gas into Nitrous Oxide, by Muriate of Tin, and dry Sulphures.*

a. Nitrous gas expofed to dry muriate of tin, is flowly converted into nitrous oxide: during this procefs the apparent diminution is to about one half; but if the products are nicely examined, and the neceffary corrections made, the real diminution of nitrous gas by muriate of tin, will be the fame as by the fulphites; that is, 100 parts of it will be converted into 48 of nitrous oxide.

During this converfion, no water is decompofed, and no nitrogene evolved. Solution of muriate of tin converts nitrous gas into nitrous oxide; but with much lefs rapidity than the folid falt.

b. Nitrous gas expofed to dry and perfectly well made fulphures, particularly fuch as are produced from cryftalifed alumn* and charcoal not fufficiently inflammable to burn in the atmofphere, is converted into nitrous oxide by the fimple abftraction of a portion of its oxygene, and confequently undergoes a diminution of $\frac{52}{100}$.

It is probable, that all the bodies having ftrong affinity for oxygene will, at certain temperatures, convert nitrous gas into nitrous oxide. Prieftley, and the Dutch chemifts, effected the change by heated fulphur. Perhaps nitrous gas fent through a tube heated, but not ignited, with phofphorus, would be converted into nitrous oxide.

IV. *Decompofition of Nitrous Gas, by Sulphurated Hydrogene.*

a. When nitrous gas and fulphurated hy-

* That is, alumn containing fulphate of potafh.

drogene are mingled together, a decompofition of them flowly takes place. The gafes are diminifhed, fulphur depofited, nitrous oxide formed, and figns of the produ&ion of ammoniac * and water perceived.

In this procefs no fulphuric, or fulphureous acid is produced; confequently none of the fulphur is oxydated, and of courfe the changes depend upon the combination of the hydrogene of the fulphurated hydrogene, with different portions of the oxygene and nitrogene of the nitrous gas, to form water and ammoniac, the remaining oxygene and nitrogene affuming the form of nitrous oxide.

This fingular exertion of attra&ions by a fimple body, appears highly improbable a priori, nor did I admit it, till the formation of ammoniac, and the non-oxygenation of the fulphur, were made evident by many experiments.

In thofe experiments, the diminution of the nitrous gas was not uniformly the fame. It

* The produ&ion of ammoniac in this procefs was ob-ferved by Kirwan and Auftin.

varied from $\frac{11}{20}$ to $\frac{14}{20}$. In the moſt accurate of them, 5 cubic inches of nitrous gas were con-verted into 2.2 of nitrous oxide. Conſequently the quantity of ammoniac formed was ,047 grains.

In experiments on the converſion of nitrous gas into nitrous oxide, by ſulphurated hydro-gene, the gaſes ſhould be rendered as dry as poſſible. The preſence of water conſiderably retards the decompoſition.

b. The ſulphures* diſſolved in water convert nitrous gas into nitrous oxide. This decom-poſition is not, however, produced by the ſimple abſtraction of oxygene from the nitrous gas to form ſulphuric acid. It depends as well on the de-

---

* Solution of ſulphure of ſtrontian, or barytes, ſhould be uſed. During the converſion of nitrous gas into nitrous oxide by thoſe bodies, a thin film is depoſited on the ſurface of the ſolution. This film examined, is found to conſiſt of ſulphur and ſulphate. Poſſibly the nitrous gas is wholly decompoſed by the hydrogene of the ſulphurated hydrogene in the ſolution, whilſt the ſulphate is produced from water decompounded by the ſulphur to form more gas for the ſaturation of the hydro-ſulphure.

compofition of the fulphurated hydrogene diffolved in the folution, or liberated from it. In this
procefs fulphur is depofited on the furface of the
fluid, fulphuric acid is formed, and the diminution, making the neceffary corrections, is nearly
the fame as when free fulphurated hydrogene
is employed.

It is extremely probable that fulphurated hydrogene, in combination with the alkalies, as
well as with water, is capable of being flowly
decompofed by nitrous gas.

V. Decompofition of Nitrous Gas by Nafcent Hydrogene.

a. When nitrous gas, is expofed to wetted
iron filings, a diminution of its volume flowly
takes place ; and after a certain time, it is found
converted into nitrous oxide.

In this procefs ammoniac* is formed, and the
iron partially oxydated.

---

* As was firft obferved by Prieftley and Auftin, and as I
have proved by many experiments.

The water in contact with the iron is decompofed by the combination of its oxygene with
that fubftance, and of its hydrogene with a portion of the oxygene and nitrogene of the nitrous
gas, to form water and ammoniac.

That the iron is not oxydated at the expence
of the oxygene of the nitrous gas, appears very
probable from the analogy between this procefs,
and the mutual decompofition of nitrous gas
and fulphurated hydrogene. Befides, dry iron
filings effect no change whatever in nitrous
gas, at common temperatures.

I have generally found about 12 of nitrous
gas converted into 5 of nitrous oxide in this
procefs; which is not very different from the
diminution by fulphurated hydrogene. It takes
place equally well in light and darknefs; but
more rapidly in warm weather than in cold.

b. Nitrous gas expofed to a large furface of
zinc, in contact with water, is flowly converted
into nitrous oxide; at the fame time that ammoniac is generated, and white oxide of zinc
formed. This procefs appears to depend, like

the laft, upon the decompofition of water by the affinities of part of the oxygene and nitrogene of nitrous gas, for its hydrogene, to form ammoniac and water; and by that of zinc for its oxygene. Zinc placed in contact with water, and confined by mercury,* decompofes it at the common temperature. Zinc, when perfectly dry, does not in the flighteft degree act upon nitrous gas.

I have not been able to determine exactly the diminution of volume of nitrous gas, during its converfion into nitrous oxide by zinc. In one experiment 20 meafures of nitrous gas, containing about ,03 nitrogene, were diminifhed to 9, after an expofure of eight days to wetted zinc; but from an accident, I was not able to afcertain the exact quantity of nitrous oxide formed.

c. It is probable that moft of the imperfect metals will be found capable of oxydation, by the decompofition of water, when its hydrogene is abftracted by the oxygene and nitrogene of

---

* As I have found by experiment.

nitrous gas. I have this day (April 14, 1800), examined two portions of nitrous gas, one of which had been expofed to copper filings, and the other to powder of tin, for twenty-three days.

The gas that had been expofed to copper was diminifhed nearly two fifths. The taper burnt in it with an enlarged flame, blue at the edges. Hence it evidently contained nitrous oxide.

The nitrous gas in contact with tin had undergone a diminution of one fourth only, and did not fupport flame.

VI. *Mifcellaneous Obfervations on the converfion of Nitrous Gas into Nitrous Oxide.*

a. Dr. Prieftley found nitrous gas expofed to a mixture of iron filings and fulphur, with water, converted after a certain time, into nitrous oxide. Sulphurated hydrogene is always pro-duced during the combination of iron and ful-phur, when they are in contact with water; and by the hydrogene of this in the nafcent

O

ftate, the nitrous gas is moft probably decompofed.

b. Green oxide of iron moiftened with water,
expofed to nitrous gas, flowly gains an orange
tinge, whilft the gas is diminifhed. . Moft
likely it is converted into nitrous oxide ; but
this I have not afcertained.

c I expofed nitrous gas, to the following bodies
over mercury for many days, without any
diminution, or apparent change in its properties.
Alcohol, faccharine matter, hydro-carbonate,
fulphureous acid, and phofphorus.

d. Cryftalifed fulphate, and muriate of iron,
abforb a fmall quantity of nitrous gas, and
become dark colored on the outfide ; but after
this abforption, (which probably depends on
their water of cryftalifation,) has taken place,
no change is effected in the gas remaining.

e The power of iron to decompofe water being
much increafed by increafe of temperature, nitrous gas is converted into nitrous oxide much
more rapidly when placed in contact with a furface of heated iron, than when expofed to it at

common temperatures. During the decompofition of nitrous gas in this way, ammoniac *
is formed.

f. The curious experiments of Rouppe,†
on the abforption of gafes by charcoal, compared with the phænomena noticed in this
Divifion, render it probable that hydrogene in
a ftate of loofe combination with charcoal, will
be found to convert nitrous gas into nitrous
oxide.

VII. *Recapitulation of conclufions concerning the converfion of Nitrous Gas into Nitrous Oxide.*

a. Certain bodies having a ftrong affinity
for oxygene, as the fulphites, dry fulphures,
muriate of tin, &c. convert nitrous gas into
nitrous oxide, by fimply attracting a portion of
its oxygene; whilft the remaining oxygene

* As was obferved by Milner. Nitrous gas paffed over
heated zinc, or tin, I doubt not will be found converted
into nitrous oxide.

† Annales de Chimie. xxxii, p. 3.

enters into combination with the nitrogene, and they assume a more condensed state of exist-ence.

b. Nitrous gas is converted into nitrous oxide by hydrogene, in a peculiar state of ex-istence, as in sulphurated hydrogene; and that by a series of very complex affinities. Both oxygene and nitrogene are attracted from the nitrous gas by the hydrogene, in such propor-tions as to form water and ammoniac, whilst the remaining oxygene and nitrogene * assume the form of nitrous oxide.

c. Nitrous gas placed in contact with bodies, such as iron and zinc decomposing water, is converted into nitrous oxide, at the same time that ammoniac is formed. It is difficult to ascertain the exact rationale of this process. For either the nascent hydrogene produced by the decomposition of the water by the metallic sub-stance may combine with portions of both the

---

* The decomposition and recomposition of water, in this process, are analogous to some of the phænomena observed by the ingenious Mrs. Fulhame.

oxygene and nitrogene of the nitrous gas ; and thus by forming water and ammoniac, convert it into nitrous oxide. Or the metallic fubftance may attract at the fame time oxygene from the water and nitrous gas, whilft the nafcent hydrogene of the water feizes upon a portion of the nitrogene of the nitrous gas to form ammoniac.

The degree of diminution, and the analogy between this procefs and the decompofition of nitrous gas by fulphurated hydrogene, render the firft opinion moft probable.

VIII. *The produ&tion of Nitrous Oxide during the oxydation of Tin, Zinc, and Iron, in Nitric Acid.*

a. **Dr.** Prieftley difcovered, that during the folution of tin, zinc, and iron, in nitric acid, certain portions of nitrous oxide were produced, mingled with quantities of nitrous gas, and nitrogene, varying in proportion as the acid employed was more or lefs concentrated.

It has long been known that ammoniac is formed during the folution of tin, zinc, and iron, in diluted nitric acid. Confequently, in thefe proceffes water is decompofed.

I had defigned to inveftigate minutely thefe phænomena, fo as to afcertain the quantities of water and acid decompounded, and of the new products generated. But after going through fome experiments on the oxydation of tin without gaining conclufive refults, the labor, and facrifice of time they demanded, obliged me to defift from purfuing the fubject, till I had completed more important inveftigations.

I fhall detail the few obfervations which have occurred to me, relating to the production of nitrous oxide from metallic folutions.

b. When tin is diffolved in concentrated nitric acid, fuch as of 1.4, nitrous oxide is produced, mingled with generally more than twice its bulk of nitrous gas. In this procefs but little free nitrogene is evolved, and the tin is chiefly precipitated in the form of a white powder. If the folution, after the generation of thefe pro-

ducts, is faturated with lime, and heated, the
ammoniacal fmell is diftinct.

When nitric acid of fpecific gravity 1.24, is
made to act upon tin; in the beginning
of the procefs, nearly equal parts of nitrous gas
and nitrous oxide are produced ; as it advances,
the proportion of nitrous oxide to the nitrous
gas increafes : the largeft quantity of nitrous
oxide that I have found in the gas procured
from tin is $\frac{3}{4}$, the remainder being nitrous gas
and nitrogene.

When tin is oxydated in an acid of lefs fpecific gravity than 1.09, the quantities of gas
difengaged are very fmall, and confift of nitrogene, mingled with minute portions of nitrous
oxide, and nitrous gas.

Whenever I have faturated folutions of tin
in nitric acid of different fpecific gravities, with
lime, and afterwards heated them, the ammoniacal fmell has been uniformly perceptible,
and generally moft diftinct when diluted acids
I have been employed.

c. When zinc is diffolved in nitric acid,

whatever is its specific gravity, certain quantities of nitrous oxide are produced.

Nitric acids of greater specific gravity than 1.2, act upon zinc with great rapidity, and great increase of temperature. The gases disengaged from these solutions consist of nitrous gas, nitrous oxide, and nitrogene; the nitrous oxide rarely equals one third of the whole.

When nitric acid of 1,104 is made to dissolve zinc, the gas obtained in the middle of the process consists chiefly of nitrous oxide. From such a solution I obtained gas which gave a residuum of one sixth only when absorbed by water. The taper burnt in it with a brilliant flame, and sulphur with a vivid rose-colored light.

100 grains of granulated zinc, during their solution in 300 grains of nitric acid, of 1,43, diluted with 14 times its weight of water, produced 26 cubic inches of gas. Of this gas $\frac{7}{36}$ were nitrous, $\frac{17}{36}$ nitrous oxide, and the remainder nitrogene. The solution saturated with lime and heated, gave a distinct smell of ammoniac.

d. During the solution of iron in concentrated nitric acid, the gas given out is chiefly nitrous ; it is however generally mingled with minute quantities of nitrous oxide. When very dilute nitric acids are made to act upon iron, by the assistance of heat, nitrous oxide is produced in considerable quantities, mingled with nitrous gas and nitrogene ; the proportions of which are smaller as the procefs advances.* The fluid remaining after the oxydation and solution of iron in nitric acid, always contains ammoniac.

e. As during the solution of tin, zinc, and iron, in nitric acid, the quantity of acid is diminished in proportion as the procefs advances, it is reafonable to suppofe that the relative quantities of the gafes evolved are perpetually varying. In the beginning of a diffolution, the nitrous gas

---

* From one of Dr. Prieftley's experiments, it appears that hydrogene gas is fometimes difengaged during the folution of iron in very dilute nitric acid by heat. This phænomenon has never occurred to me.

generally predominates, in the middle nitrous oxide, and at the end nitrogene.

f. During the generation of nitrous gas, nitrous oxide, and ammoniac, from the decompofition of folution of nitric acid in water, by tin, zinc, and iron, very complex attractions muft exift between the conftituents of the fubftances employed. The acid and the water are decompofed at the fame time, and in proportions different as the folution is more concentrated, by the combination of their oxygene with the metallic body.

The nitrous gas is produced by the combination of the metal with $\frac{32}{100}$ of the oxygene of the acid. The nitrous oxide is moft probably generated by the decompofition of a portion of the nitrous gas difengaged, by the nafcent hydrogene of the water decompounded; fome of it may be poffibly formed from a more complete decompofition of the acid.

The production of ammoniac may arife, probably from two caufes; from the decompofition of the nitrous gas by the combination of the nafcent hydrogene of the water, with portions of its oxygene and nitrogene at the fame time ; and from the union of hy- drogene with nafcent nitrogene liberated in confequence of a complete decompofition of part of the acid.

IX. *Additional Obfervations on the pro- duction of Nitrous Oxide.*

a. When nitric acid is combined with mu- riatic acid, or fulphuric acid,* the quantities of nitrous oxide produced from its decompofition by tin, zinc, and iron, are rather increafed than diminifhed. The nitrous oxide obtained from thefe folutions is, however, never fufficiently pure for phyfiological experiments. It is always mingled with either nitrous gas, nitrogene, or hydrogene, and fometimes with all of them.

b. From the folutions of bifmuth, nickel,

* As was difcovered by Prieftley, and the Dutch Che- mifts.

lead, and copper, in diluted nitric acid, I have never obtained any perceptible quantity of nitrous oxide : the gas produced is nitrous, mingled with different portions of nitrogene. Antimony and mercury, during their folution in aqua regia, give out only nitrous gas.

Probably none of the metallic bodies, except thofe that decompofe water at temperatures below ignition, will generate nitrous oxide from nitric acid. On cobalt and manganefe I have never had an opportunity of experimenting : manganefe will probably produce nitrous oxide.

c. During the folution of vegetable matters* in nitric acid, by heat, very minute portions of nitrous oxide are fometimes produced, always however mingled with large quantities of nitrous gas, and carbonic acid.

When nitric acid is decompounded by ether, fixed oils, volatile oils, or alcohol, towards the end of the procefs fmall quantities of nitrous oxide are produced, and

* Such as the leaves, bark, and wood, of treee.

fometimes fufficiently pure to fupport the flame of the taper.†

d. When green oxide of iron is diffolved in nitric acid, nitrous oxide is produced, mingled with nitrogene and nitrous gas.

e During the converfion of green fulphate, or green muriate of iron into red, by the decompofition of dilute nitric acid, nitrous oxide is formed, mingled with different proportions of nitrous gas and nitrogene.

f. When folution of green nitrate of iron is heated, a part of the acid is decompofed, red oxide is precipitated, red nitrate formed, and impure nitrous oxide evolved.

g. When iron is introduced into a folution of nitrate of copper, the copper is precipitated in its metallic ftate, whilft nitrous oxide, mingled with fmall portions of nitrogene, is produced.* Both zinc and tin precipitate copper in its metallic form from folution in the nitric acid.

† As I have obferved, after Prieftley.

* As was difcovered by Prieftley.

During thefe precipitations, certain quantities of nitrous oxide are generated, mingled however with larger quantities of nitrogene than that produced from decompofition by iron. In all thefe proceffes ammoniac is formed, and water confequently decompofed.

The decompofition of water and nitric acid, during the precipitation of copper from folution of nitrate of copper, by tin, zinc, and iron, depends upon the ftrong affinity of thofe metals for oxygene, and their powers of combining with a larger quantity of it than copper.

X. *Decompofition of Aqua Regia by Platina, and evolution of a Gas analogous to Oxygenated Muriatic Acid, and Nitrogene.*

a. De la Metherie, in his effay on different airs, has afferted that the gas produced by the folution of platina in nitro-muriatic acid, is identical with the dephlogifticated nitrous gas of Prieftley. He calls it nitrous gas with excefs of pure air, and affirms that it diminishes, both with nitrous gas and common air.

b. I introduced into a veffel containing 30 grains of platina, 2050 grains of aqua regia, compofed of equal parts, by weight, of concentrated nitric acid of 1,43, and muriatic acid of 1,16. At the common temperature, that is, 49°, no action between the acid and platina appeared to take place. On the application of the heat of a fpirit lamp, the folution gradually became yellow red, and gas was given out with rapidity. Some of this gas received in a jar filled with warm water, appeared of a bright yellow color. On agitation, the greater part of it was abforbed by the water, and the remainder extinguifhed flame. When it was received over mercury, it acted upon it with great rapidity, and formed on the furface a white cruft.

As the procefs of folution advanced, the color of the acid changed to dark red, at the fame time that the production of gas was much increafed; more than 40 cubic inches were foon collected in the water apparatus.

6. Different portions of the gas, were examined, it exhibited the following properties :

1. Its color was orange red,* and its smell exactly resembled that of oxygenated muriatic acid.

2. When agitated in boiled water, it was rapidly absorbed, leaving a residuum of rather more than one twelfth.

3. The taper burnt in it with increased brilliancy, the flame being long, and deep red at the edges.

4. Iron introduced into it ignited, burnt with a dull red light.

5. Green vegetables exposed to it were instantly rendered white.

6. It underwent no diminution, mingled with atmospheric air.

7. When mingled with nitrous gas, it gave dense red vapor, and rapid diminution.

---

* This deep color depended, in some measure, upon the nitro-muriatic vapor suspended in it. I have since observed that it is more intense in proportion as the heat employed for the production of the gas has been stronger. The natural color of the peculiar gas is deep yellow.

c. From the exhibition of thefe properties, it was evident that the gas produced during the folution of platina in aqua regia, chiefly confifted of oxygenated muriatic acid, or of a gas highly analogous to it. It was, however, difficult to conceive how a body, by combining with a portion of the oxygene of nitro-muriatic acid, could produce from it oxygenated muriatic acid, apparently mingled with very fmall portions of any other gas.

d. To afcertain whether any permanent gas was produced during the ebullition of aqua regia, of the fame compofition as that ufed for the folution of the platina ; I kept a large quantity of it boiling for fome time, in communication with the water apparatus ; the gas generated appeared to be wholly nitro-muriatic, and was abforbed as faft as produced, by the water.

e. To determine whether any nitrous oxide was mingled with the peculiar gas, as well as the nature and quantity of the unabforbable gas, nitrous gas was gradually added to 21 cubic inches of the gas produced from a new folution,

till the diminution was complete : the gas remaining equalled 2,3 cubic inches ; it was
unabforbable by water, and extinguifhed flame.

In another experiment, when the the laft
portions of gas from a folution were carefully
received in water previoufly-boiled, 12 cubic
inches agitated in water left a refiduum of 1,3 ;
whilft the fame quantity decompofed by nitrous
gas, containing ,02 nitrogene, left about 1.5

Hence it appeared that the aëriform products
of the folution confifted of the peculiar gas
analogous to oxygenated muriatic acid, and of
a fmall quantity of nitrogene.

f. Confequently a portion of the nitric acid
of the aqua regia had been decompofed ; but if
it had given oxygene both to the platina and
muriatic acid, the quantity of nitrogene evolved
ought to have been much more confiderable.

g. To afcertain if any water had been decompofed, and the nitrogene condenfed in the
folution by its hydrogene, to form ammoniac,
I faturated a folution with lime, and heated it,
but no ammoniacal fmell was perceived.

b. To determine if any nitrogene had entered into chemical combination with muriatic acid and oxygene, fo as to form an aëriform triple compound, analogous in its properties to oxygenated muriatic acid, I expofed fome of the gas to mercury, expecting that this fubftance, by combining with its oxygene, would effect a complete decompofition; and this was actually the cafe: for the gas was at firft rapidly diminifhed; and the mercury became oxydated; its volume, however, foon increafed; and the refidual gas appeared to be nitrous, mingled with much nitrogene. The exact proportions of each, from an accident, I could not determine.

This experiment was inconclufive, becaufe the nitro-muriatic acid fufpended in the peculiar gas, from which it can probably be never perfectly freed, acted in common with it upon the mercury, and produced nitrous gas : and this nitrous gas, at the moment of its production, formed nitrous acid by combining with the oxygene of the peculiar gas; and the nitrous 'acid generated * was again decompofed by the mercury; and hence nitrous gas evolved, and poffibly fome nitrogene.

i. Peculiar circumftances prevented me at this time from completely inveftigating the fubject. It remains doubtful whether the gas confifts fimply of highly oxygenated muriatic acid and nitrogene,† produced by the decom-

---

* The decompofition of aëriform nitrous acid by mercury, was firft noted by Prieftley; vol. iii. pag. 101. This decompofition I have often had occafion to obferve. In reading Humbolt's paper on eudiometry, Annales de Chimie, xxviii, pag. 150, I was not a little furprifed to find that he takes no notice of this fact. He feems to fuppofe that nitrous acid can remain aëriform, and even be condenfed, in contact with mercury, without alteration. He fays, "In mingling 100 parts of atmofpheric air with 100 of nitrous air, the air immediately became red, but all the acid produced remained aëriform; and after eighteen hours fome *drops* only of acid were formed, which *fwam* upon the mercury."

† Lavoifier has faid concerning aqua regia, "In folu-" tions of metals in this acid, as in all other acids, the " metals are firft oxydated, by attracting a part of the " oxygene from the compound radical. This occafions the " difengagement of a particular fpecies of gas not hitherto " defcribed, which may be called nitro-muriatic gas. It

-pofition of nitric acid from the coalefcing affinities of platina and muriatic acid for oxygene ;
or whether it is compofed of a *peculiar* gas,
analogous to oxygenated muriatic acid, and
nitrogene, generated from fome unknown
affinities.†

XI. *On the action of the Electric Spark on a
mixture of Nitrogene and Nitrous Gas.*

Thinking it poffible that nitrous gas and

" has a very difagreeable fmell, and is fatal to animal life
" when refpired ; it attacks iron, and caufes it to ruft ; it
" is abforbed in confiderable quantities by water." Elem.
Eng. 237.

† I have no doubt but that the gas procured from the
folution of gold in aqua regia, is analogous to that produced from platina.

Some very uncommon circumftances are attendant on the
folution of platina :

1ft. The immenfe quantity of acid required for the folution of a minute quantity of platina.

2d. The great quantity of gas produced during the folution of this minute quantity.

3d. The intenfe red color of the folution, and its perfectly acid properties after it ceafes to act upon the metal.

nitrogene might be made to combine by the action of the electric fpark, fo as to form nitrous oxide, I introduced 20 grain meafures of each of them into a fmall detonating tube, graduated to grains, ftanding over mercury, and containing a very fmall quantity of cabbage Juice rendered green by an alkali. After electric fparks had been paffed through the gafes for an hour and half, they were diminifhed to about 32, and the cabbage juice was flightly reddened. On introducing about 10 meafures of hydrogene, and paffing the electric fpark through the whole, no inflammation or diminution was perceptible. Hence the condenfation moft probably arofe wholly from the formation of nitrous acid,* by the more intimate union of the oxygene of nitrous gas with fome of its nitrogene; as in the experiments of Prieftley.

As the nafcent nitrogene, in the decompo-

---

* For if nitrous oxide had been formed, it would have been decompofed by the hydrogene.

fition of nitrate of ammoniac, combines with a portion of oxygene and nitrogene, to form nitrous oxide ; it is probable that nitrous oxide may be produced during the paffage of nitrous gas and ammoniac through a heated tube.

XII. *General Remarks.*

There are no reafons, for fuppofing that nitrous oxide is formed in any of the proceffes of nature ; and the nice equilibrium of affinity by which it is conftituted, forbids us to hope for the power of compofing it from its fimple princi-ples. We muft be content to produce it, either directly or indirectly, from the decom-pofition of nitric acid. And as in the decom-pofition of nitrate of ammoniac, not only all the nitrogene of the nitric acid enters into the com-pofition of the nitrous oxide produced, but likewife that of the ammoniac, this procefs is by far the cheapeft, as well as the moft expe-ditious. A mode of producing ammoniac at little expence, has been propofed by Mr. Watt. Condenfed in the fulphuric acid, it can be eafily made to combine with nitric acid, from the decompofition of nitre by double affinity. And thus, if the hopes which the experiments at the end of thofe refearches induce us to indulge, do not prove fallacious, a fubftance which has been heretofore almoft exclufively appropriated to the deftruction of mankind, may become, in the hands of philofophy, a means of producing health and pleafurable fenfation.

RESEARCH II.

INTO THE COMBINATIONS OF

NITROUS OXIDE,

AND ITS

DECOMPOSITION

BY

COMBUSTIBLE BODIES.

RESEARCH II.

DIVISION I.

EXPERIMENTS and OBSERVATIOUS on the COMBINATIONS of NITROUS OXIDE.

—————

1. *Combination of Water with Nitrous Oxide.*

a. THE difcoverer of nitrous oxide firft obferved its folubility in water; and it has fince been noticed by different experimentalifts.

Dr. Prieftley found that water diffolved about one half of its bulk of nitrous oxide, and that at the temperature of ebullition, this fubftance was incapable of remaining in combination with it.*

* Experiments and obfervations, vol. ii. pag. 81.

b. I introduced to 9 cubic inches of pure water, i. e. water diftilled under mercury, 7 cubic inches of nitrous oxide, which had been obtained over mercury, from the decompofition of nitrate of ammoniac, and in confequence was perfectly pure. After they had remained together for 11 hours, temperature being 46°, during which time they were frequently agitated, the gas remaining was 2,3; confequently 4,7 cubic inches had been abforbed. And then, 100 cubic inches, $= 25300$ grains of water, will abforb 54 cubic inches, $= 27$ grains, of nitrous oxide.

c. The tafte of water impregnated with nitrous oxide, is diftinctly fweetifh; it is fofter than common water, and, in my opinion, much more agreeable to the palate. It produces no alteration in vegetable blues, and effects no change of color in metallic folutions.

d. Thinking that water impregnated with nitrous oxide might probably produce fome effects when taken into the ftomach, by giving out its gas, I drank, in June, 1799, about

3 ounces of it, but without perceiving any effects.

A few days ago, confidering this quantity as inadequate, I took at two draughts nearly a pint, fully faturated ; and at this time Mr. Jofeph Prieftley drank the fame quantity.

We neither of us perceived any remarkable effects.

Since that time I have drank near three pints of it in the courfe of a day. In this inftance it appeared to act as a diuretic, and I imagined that it expedited digeftion. As a matter of tafte, 1 fhould always prefer it to common water.

e. Two cubic inches of pure water, that had been made to abforb about 1,1 cubic inches of nitrous oxide ; when kept for fome time in ebullition, and then rapidly cooled, produced nearly .1 of gas. Sulphur burnt in this gas with a vivid rofe-colored flame.

In another experiment, in which the gas was expelled by heat from impregnated water, and abforbed again after much agitation on cooling ; the refiduum was hardly perceptible, and moft likely depended upon fome gas which had adhered to the mercury, and was liberated during the ebullition. Hence it appears that nitrous gas is expelled unaltered from its aqueous folution by heat.

f. `I have before mentioned, Divifion III, that nitrous oxide, during its combination with fpring water, expels the common air diffolved in it. This common air generally amounts to one fixteenth, the volume of the water being unity. A correction on account of this circumftance muft be made for the apparent deficiency of diminution, and for the common air mingled in confequence, with nitrous oxide during its abforption by common water.

g. Water impregnated with nitrous gas abforbed nitrous oxide ; but the refidual gas was much greater than that of common water, and gave red fumes with atmofpheric air. Nitrous gas agitated for a long while over water highly impregnated with nitrous oxide, was not in the flighteft degree diminifhed, in one experiment indeed it was rather increafed ; doubtlefs from the liberation of fome nitrous oxide from the water by the agitation.

h. Nitrous oxide kept in contact with aqueous folution of fulphurated hydrogene and often agitated, was not in the flighteft degree diminifhed.

Sulphurated hydrogene, introduced into a folution of nitrous oxide, was rapidly abforbed, and as the procefs advanced, the nitrous oxide was given out.

i. Water impregnated with carbonic acid, poffeffed no action upon nitrous oxide, and did not in the flighteft degree abforb it. When carbonic acid was introduced to an aqueous folution of nitrous oxide ; the aëriform acid was abforbed, and the nitrous oxide liberated.

k. From thefe obfervations it appears that nitrous oxide has lefs affinity for water, than even the weaker acids, fulphurated hydrogene and carbonic acid ; as indeed one might have conjectured a priori from its degree of folubility : likewife that it has a ftronger attraction for water than the gafes not poffeffed of acid or alkaline properties ; it expelling from water nitrous gas, oxygene, and common air ; probably hydro-carbonate, hydrogene, and nitrogene.

II. *Combinations of Nitrous Oxide with Fluid Inflammable Bodies.*

a. Vitriolic ether abforbs nitrous oxide in much larger quantities than water.

A cubic inch of ether, at temperature 52°, combined with a cubic inch and feven tenths of nitrous oxide.

Ether thus impregnated was not at all altered in its appearance ; its fmell was precifely the fame, but the tafte appeared lefs pungent, and more agreeable. Nitrous oxide is liberated unaltered from ether at a very low temperature, that is, at about the boiling point of this fluid.

For expelling nitrous oxide from impregnated ether, and for afcertaining in general the quantity of gafes combined with fluids, I have lately made ufe of a very fimple method, which it may not be amifs to defcribe.

The impregnated fluid is introduced into a fmall thin tube, graduated to ,05 cubic inches, through mercury. The quantity of fluid fhould never equal more than a fifth or fixth of the capacity of the tube.

The lower part of the tube is adapted to an orifice in the fhelf of the mercurial apparatus, fo as to make an angle of about 40° with the furface of the mercury.

The flame of a fmall fpirit lamp is then applied to that part of the tube containing the fluid; and after the expulfion of the gas from it, the heat is raifed fo as to drive out the fluid through the orifice of the tube. Thus the liberated gas is preferved in a ftate proper for accurate examination.

Impregnated ether, during its combination with water, gives out the greater part of its nitrous oxide. During the liberation of nitrous oxide from ether, by its combination with water, a very curious phænomenon takes place.

If the water employed is colored, fo that it may be feen in a ftratum diftinct from the im-

Q

pregnated ether, at the point of contact a number of fmall fpherules of fluid will be perceived, apparently repulfive both to water and ether; thefe fpherules become gradually covered with minute globules of gas, and as this gas is liberated from their furfaces, they gradually difappear.

b. Alcohol diffolves confiderable quantities of nitrous oxide.

2 cubic inches of alcohol, at 52°, combined with 2,4 cubic inches of nitrous oxide. The alcohol thus impregnated had a tafte rather fweeter than before, but in other phyfical properties was not perceptibly altered.

Nitrous oxide is incapable of remaining in combination with this fluid at the temperature of ebullition; it is liberated from it unaltered by heat.

Impregnated alcohol, during its combination with water, gives out the greater part of its combined nitrous oxide : on mingling the two fluids together, at the point of contact the alcohol becomes covered with an infinite number of fmall globules of gas, which continue to be generated during the whole of the combination, and in paffing through the fluid render it almoft opaque.

c. The effential, oils abforb nitrous oxide to a greater extent than either alcohol or ether.

,5 cubic inches of oil of carui combined with 1,2 cubic inches of nitrous oxide at 51°. The color of the oil thus impregnated was rather paler than before.

Nitrous oxide is expelled unaltered from impregnated oil of carui, by heat.

,1 of oil of turpentine abforbed nearly 2 of nitrous oxide, at 57°. Its properties were not fenfibly altered from this combination, and the gas was expelled from it undecompounded, by heat.

d. As well as the effential oils, the fixed oils diffolve nitrous oxide at low temperatures, whilft at high temperatures they do not remain in combination.

,1 of olive oil abforbed, at 61°, 1,2 of nitrous oxide, but without undergoing any apparent phyfical change.

III. *Action of Fluid Acids on Nitrous Oxide.*

a. Nitrous oxide expofed to concentrated fulphuric acid, undergoes no change, and fuffers no diminution, that may not be accounted for from the abftraction of a portion of its water by the acid.

b. Nitrous oxide is fcarcely at all foluble in nitrous acid, and expofed to that fubftance, undergoes no alteration.

c. Muriatic acid, of fpecific gravity 1,14 abforbs about a third of its bulk of nitrous oxide. It fuffers no apparent change in its properties from being thus impregnated, and the gas is again given out from it on the application of heat.

d. Acetic acid abforbs nearly one third of its bulk of nitrous oxide.

e. Aqua regia, that is, the nitro-muriatic acid, abforbs a very minute portion of nitrous oxide.

f. Nitrous oxide was expofed to a new compound acid, confifting of oxygenated muriatic acid, and fulphuric acid, which I difcovered in

July, 1799, and of which an account will be
fhortly publifhed; but it was neither abforbed or
altered.

I have before mentioned that the aqueous
folutions of fulphurated hydrogene and carbonic
acid, neither diffolve or alter nitrous oxide.

IV. *Action of Saline Solutions, and other
Subftances, on Nitrous Oxide.*

a. Nitrous oxide expofed to concentrated
folution of green fulphate of iron, at 58°, underwent no perceptible diminution; not even
after it had been fuffered to remain in contact
with it for half an hour.

b. It underwent diminution of nearly 2 when
agitated in contact with a folution of red fulphate of iron, the volume of the folution being
unity.

c. Solution of green fulphate of iron, fully
impregnated with nitrous gas, did not in the
flighteft degree abforb nitrous oxide, and appeared to have no action upon it.

d. Solution of green muriate of iron, whether impregnated with nitrous gas, or unimpregnated, has no affinity for, or action upon, nitrous oxide.

e. Solution of red muriate of iron in alcohol, abforbed nearly one fifth of its bulk, of nitrous oxide.

f. Solution of pruffiate of potafh abforbed nearly one third of its volume, of nitrous oxide, which was again expelled from it by heat.

g. Solution of nitrate of copper appeared to have no affinity for nitrous oxide.

h. Concentrated folution of nitrate of ammoniac, at 58°, abforbed one eighth of its bulk of nitrous oxide.

i. Solutions of alkaline fulphures abforb nitrous oxide in quantities proportionable to the water they contain; it is expelled from them unaltered by heat. None of the hydrofulphures diffolve more than half their bulk of nitrous oxide.

k. Concentrated folutions of the fulphites poffefs little or no action on nitrous oxide :

diluted folutions abforb it in fmall quantities.

l. Concentrated folution of muriate of tin abforbs about one eighth of nitrous oxide; more dilute folutions abforb larger quantities.

From thefe obfervations we learn, that neutro-faline folutions in general, have very feeble attractions for nitrous oxide; and as folutions of green muriate, and fulphate of iron, whether free from nitrous gas, or impregnated with it, poffefs no action upon nitrous oxide, nitrous gas may be feparated from this fubftance by thofe folutions with greater facility than nitrous oxide can be feparated from nitrous gas, by water or alcohol. Charcoal abforbs nitrous oxide as well as all other gafes; and it is difengaged from it by heat.

I have as yet found no other folid body, not poffeffed of alkaline properties, capable of ab-forbing nitrous oxide in any ftate of exiftence.

The bodies poffeffing the ftrongeft affinity for oxygene, the dry fulphites, muriate of tin, the common fulphures, white pruffiate of potafh, and green oxide of iron, do not in the flighteft

degree, act on nitrous oxide at common temperatures.

V. *Action of different Gases on Nitrous
Oxide.*

a. 12 measures of muriatic acid gas were mingled with 7 measures of nitrous oxide at 56°. After
remaining together, for a minute, they filled a
space equal to $19\frac{1}{2}$ measures. When water
was introduced to them, the muriatic acid was
absorbed much more slowly than if it had been
unmingled.

In another experiment, when the gases were
saturated with water, 9 measures of each of
them, when mingled and suffered to remain in
contact for a quarter of an hour, filled a space
nearly equal to 19; and after the muriatic acid
had been absorbed by potash, the nitrous oxide
remained unaltered in its properties.

From the expansion, it appears most probable that aëriform muriatic acid, and nitrous
oxide, have a certain affinity for each other, and

that they combine when mingled together; for in the laft, experiment, the increafe of volume cannot be accounted for by fuppofing that nitrous oxide undergoes lefs change of volume than muriatic acid, by aëriform combination with water, and that the expanfion depended upon the folution of fome of its combined water by the muriatic acid. That muriatic acid and nitrous oxide have a flight affinity for each other, likewife appears from the abforption of nitrous oxide by aqueous folution of muriatic acid.

Thinking that nitrous oxide might attract muriatic acid from its folution in water, I expofed a minute quantity of fluid muriatic acid to nitrous oxide; but no alteration of volume took place in the gas.

b. 6 meafures of nitrous oxide were mingled with 11 meafures of fulphureous acid, faturated with water; after remaining at reft for fix minutes, they filled a fpace nearly equal to 18 meafures. Expofed to water, the fulphureous acid was abforbed, but not nearly fo rapidly as when in a free state. Sulphur burnt with a vivid flame in the refidual nitrous oxide. 7 meafures of fulphureous acid were now mingled with 8 of nitrous oxide. They filled a fpace nearly equal to 15¾, and no farther expanfion took place afterwards.

From thefe experiments it appears probable that fulphureous acid, and nitrous oxide, have fome affinity for each other.

c. 11 meafures of carbonic acid were mingled with 8 of nitrous oxide; they filled a fpace nearly equal to 19 meafures. On expofing the mixture to cauftic potafh, the carbonic acid was abforbed, and the nitrous oxide remained pure. Hence it appears that carbonic acid and nitrous oxide do not combine with each other.

d Oxygenated muriatic acid, and nitrous oxide, were mingled in a water apparatus: there was a flight appearance of condenfation; but this was moft probably owing to abforption by the water; on agitation, the oxygenated muriatic acid was abforbed, and the greater part of the nitrous oxide remained unaltered.

e. Sulphurated hydrogene and nitrous oxide, mingled together, neither expanded or contracted ; expofed to folution of potafh, the acid* only was abforbed.

f. 10 meafures of nitrous gas were admitted to 12 of nitrous oxide at 59°. They filled a fpace equal to 22, and after remaining together for an hour, had undergone no change. Solution of muriate of iron abforbed the nitrous gas without affecting the nitrous oxide.

g. Nitrous oxide was fucceffively mingled with oxygene, atmofpheric air, hydro-carbonate, phofphorated hydrogene, hydrogene, and nitrogene, at 57° ; it appeared to poffefs no action on any of them, and was feparated by water, the gafes remaining unaltered.

h. As nitrous oxide was foluble in ether, alcohol, and the other inflammable fluids, it was reafonable to fuppofe that its affinity for thofe bodies would enable them to unite with

* The experiments of Berthollet have clearly proved the perfect acidity of this fubftance.

it in the aëriform ftate. At the fuggeftion of
Dr. Beddoes I made the following experiment:

To 12 meafures of nitrous oxide, at 54°, I
introduced a fingle drop of ether; the gas immediately began to expand, and in four minutes
filled a fpace equal to fixteen meafures and a
quarter. When an inflamed taper was plunged
into the gas thus holding ether in folution, a
light blue flame flowly paffed through it.

A confiderable diminution of temperature is
moft probably produced, from the great expanfion of nitrous oxide during its combination
with ether.

A drop of alcohol was admitted to 14 meafures of nitrous oxide. In five minutes, the
gas filled a fpace equal to fifteen and a third;
but no farther diminution took place afterwards.

A minute quantity of oil of turpentine was
introduced to 14 meafures of nitrous oxide; it
filled, in 4 minutes, a fpace rather lefs than 14;
and no farther change took place afterwards.
Moft likely this contraction arofe from the precipitation of the water diffolved in the gas by

the ſtronger affinity of the oil for nitrous oxide. To aſcertain with certainty if any oil had been diſſolved by the gas, I introduced into it a ſmall quantity of ammoniac. It immediately became ſlightly clouded, moſt probably from the formation of ſoap, by the combination of the diſſolved oil with the ammoniac.

From theſe experiments we learn, that when nitrous oxide is mingled with either carbonic acid, oxygene, common air, hydro-carbonate, ſulphurated hydrogene, hydrogene, or nitrogene, they may be ſeparated from each other without making any allowance for contraction or expanſion: but if a mixture of either muriatic acid, or ſulphureous acid gas, with nitrous oxide, is experimented upon ; in the abſorption of the acid by alkalies, the apparent volume of gas condenſed will be leſs than the real one, by a quantity equal to the ſum of expanſion from combination. Conſequently a correction muſt be made on account of this circumſtance.

Though alcohol, ether, eſſential oils, and the fluid inflammable bodies in general, diſſolve nitrous oxide with much greater rapidity than water, yet as we are not perfectly acquainted with their action on unabforbable gafes, it is better to employ water for feparating nitrous oxide from thefe fubftances; particularly as that fluid is more or lefs combined with all gafes, and as we are acquainted with the extent of its action upon them.

By purfuing the fubject of the folution of effential oils in gafes, we may probably difcover a mode of obtaining them in a ftate of abfolute drynefs. For if other gafes as well as nitrous oxide, have a ftronger affinity for oils than for water, water moft probably will be precipitated from them during their folution of oils; and after their faturation with oil, it is likely that they are capable of being deprived of that fub- ftance by ammoniac.

VI. *Action of aëriform Nitrous Oxide on the Alkalies. History of the discovery of the combina- tions of Nitrous Oxide with the Alkalies.*

a. When nitrous oxide in a free ftate is

expofed to the folid cauftic alkalies and alkaline earths, at common températures, it is neither abforbed nor acted upon ; when it is placed
in contact with folutions of them in water; a
fmall quantity is diffolved; but this combination appears to depend on the water of the folution, for the gas can be expelled unaltered, at
the temperature of ebullition.

b. Cauftic potafh was expofed to nitrous
oxide for 13 hours : the diminution was not to
one fiftieth, and this flight condenfation moft
probably depended upon its combination with
the water of the gas.

Concentrated folution of potafh abforbed
a fourth of its bulk of nitrous oxide. When
the impregnated folution was heated, globules
of gas were given out from it rapidly ; but the
quantity collected was too fmall to examine.

Soda, whether folid or in folution, exhibited
exactly the fame phænomena with nitrous oxide.
The folution of foda abforbed near a quarter of
its bulk of gas.

c. 11 meafures of ammoniacal gas were

mingled with 8 meafures of nitrous oxide over
dry mercury, both of the gafes being faturated
with water. No change of appearance was
produced by the mixture, and they filled, after
two minutes, a fpace equal to 19. On the introduction of a little water, the ammoniac was
abforbed, and the nitrous oxide remained unaltered, for it was diffolved by water as rapidly
as if it had never been mingled with ammoniac.*

7 meafures of nitrous oxide, expofed to 6
meafures of folution of ammoniac in water, was
in an hour diminifhed to $4\frac{1}{2}$ nearly. When
the folution was heated over mercury, permanent
gas was produced, which was unabforbable by
a minute quantity of water, and foluble in a
large quantity; confequently it was nitrous
oxide.

---

* The Dutch chemifts have afferted, that mixture with
ammoniac prevents the abforption of nitrous oxide by water, either wholly or partially. Journal de Phyfique,
t. xliii. part ii. pag. 327. It is difficult to account for
their miftake.

d. Nitrous oxide was expofed to dry cauftic ftrontian; it underwent a diminution of nearly one fortieth, which moft likely was owing to the combination of the ftrontian with its water.

11 meafures of nitrous oxide were agitated in contact with 8 of ftrontian lime water: nearly 4 meafures were abforbed. The impregnated folution expofed to heat, rapidly gave out its gas; 3 meafures were foon collected, which mingled with a fmall quantity of hydrogene, and inflamed by the taper, gave a fmart detonation.

e. Nitrous oxide expofed to lime and argil, both wet and dry, was not in the flighteft degree acted upon.

From thefe experiments it is evident that nitrous oxide in the aëriform ftate cannot be combined either with the alkalies, or the alkaline earths. That a combination may be effected between nitrous oxide and thefe fubftances, it muft be prefented to them, in the *nafcent ftate.* The falts compofed of the alkalies and nitrous oxide, are not analogous to any other compound

R

fubſtances, being poſſeſſed of very ſingular properties. Before theſe properties are detailed, it
may not be amiſs to give an account of the
accidental way in which I diſcovered the mode
of combination.

In December, 1799, deſigning to make a
very delicate experiment, with a view to
aſcertain if any water was decompoſed during the converſion of nitrous gas into nitrous
oxide, by ſulphite of potaſh, I expoſed 200
grains of cryſtaliſed ſulphite of potaſh, containing great ſuperabundance of alkali, to 14
cubic inches of nitrous gas, containing one
eighteenth nitrogene. The alkali was employed to preſerve any ammoniac that might be
formed, in the free ſtate, as it would otherwiſe
combine with ſulphureous acid.*

The volume of gas diminiſhed with great
rapidity; in two hours and ten minutes it was

---

* Sulphureous acid ſaturates more potaſh than ſulphuric
acid, ſo that moſt probably during the converſion of ſulphite of potaſh into ſulphate, portions of ſulphureous acid
are diſengaged.

reduced to $6\frac{4}{5}$, which I confidered as the limit of diminution. Accidentally, however, fuffering it to remain for three hours longer, I was much furprifed by finding that not quite 12 cubic inches remained, which confifted of, nitrous oxide, mingled with the nitrogene that exifted before the experiment.

In accounting theoretically for this phænomenon, different fuppofitions neceffarily prefented themfelves.

1ft, It was poffible, that though fulphite of potafh, and potafh, feparately poffeffed no action on free nitrous oxide, yet in combination they might exert fuch affinities upon it as either to abforb it, or make it enter into new combinations.

2dly. It was more probable that the cauftic potafh, though incapable of condenfing aëriform nitrous oxide, was yet poffeffed of a ftrong affinity for it when in the *nafcent ftate*, and that the nitrous oxide condenfed in the experiment had been combined in this ftate with the free alkali.

To afcertain if the compound of potafh and

fulphite of potafh with fulphate, was capable of acting upon nitrous oxide, I fuffered a quantity of this fubftance to remain in contact with the gas, for near a day : no change whatever took place.

To determine whether the diminution of nitrous oxide depended upon its abforption in the nafcent ftate, by the peculiar compound of potafh and fulphite of potafh, or if it was fimply owing to the alkali.

I mingled a folution of fulphite of potafh with cauftic foda ; the falt, after being evaporated at a low temperature, was expofed to nitrous gas. The nitrous oxide formed was abforbed, but in rather lefs quantities than when alkaline fulphite of potafh was employed.

Hence it was evident that the alkali was the agent that had condenfed the nitrous oxide in thofe experiments, for foda is incapable of combining either with fulphate, or fulphite of potafh.

To afcertain whether any change in the conftitution of the nitrous oxide had been produced

by the condenfation, I introduced a fmall quantity of fulphite of potafh, with excefs of alkali,
that had abforbed nitrous oxide, into a long
and thin cylindrical tube filled with mercury ;
and inclining it at an angle of 35° with the
plane of the mercury, applied the heat of a fpirit
lamp to that part of the tube containing the
falts ; when the glafs became very hot, gas was
given out with rapidity ; in lefs than a minute
the tube was full. · This gas was transfered into
another tube, and examined ; it proved to be
nitrous oxide in its higheft ftate of purity ;* for
a portion of it abforbed by common water, left
no more than a refiduum of $\frac{1}{15}$, and fulphur
burnt in it with a vivid rofe-colored flame.

·Being now fatisfied that the alkalies were
capable of combining with nitrous oxide ; to
inveftigate with precifion the nature of thefe
new compounds, I proceeded in the following
manner.

* Hence we learn that fulphite of potafh, when ftrongly
heated, does not decompofe nitrous oxide, even in
the *nafcent ftate*.

VII. *Combination of Nitrous Oxide with Potaſh.*

a. Into a ſolution of ſulphite of potaſh, which had been made by paſſing ſulphureous acid gas from a mercurial airholder into cauſtic potaſh diſſolved in water, I introduced 17 grains of dry potaſh. The whole evaporated at a low temperature, gave 143 grains of ſalt. This ſalt was not *wholly* compoſed of ſulphite of potaſh and potaſh ; it contained as well, a minute quantity of carbonate, and ſulphate of potaſh, formed during the evaporation.*

120 grains of it finely pulveriſed, and retaining the water of cryſtaliſation, were expoſed to 15 cubic inches of nitrous gas, over mercury. The nitrous gas diminiſhed with great rapidity, and in three hours a cubic inch and nine tenths

---

* See the excellent memoir of Fourcroy and Vauquelin on the ſulphureous acid, and its combinations. Annales de Chimie, ii, 54. Or Nicholſon's Phil. Journal, vol. i, pag. 313.

only remained, which confifted of nearly one third nitrous oxide, and two thirds nitrogene that had pre-exifted in the nitrous gas. The increafe of weight of the falt could not be determined, as fome of it was loft by adhering to the veffel in which the combination was effected, and to the mercury. It prefented no diftinct feries of cryftalifations, even when examined by the magnifier; rendered green vegetable blues, and its tafte was very different from that of the remaining quantity of falt that had been expofed to the atmofphere. A portion of it ftrongly heated over mercury, gave out gas with great rapidity, which had all the properties of the pureft nitrous oxide.

When water was poured upon fome of it, no gas was given out, and the whole was equably and gradually diffolved. Alcohol, as well as ether, appeared incapable of diffolving any part of it.

When muriatic acid was introduced into it, confined by mercury, a rapid effervefcence took place. Part of the gas difengaged was fulphureous acid, and carbonic acid ; the remainder was nitrous oxide.

b. I made a number of experiments upon falts procured in the manner I have juft defcribed, with a view to obtain the compound of nitrous oxide and potafh, free from admixture of other falts.

When the mixed falt was boiled in alcohol or ether, no part of it appeared to be diffolved. Finding that little or no gas was given out during the ebullition of concentrated folutions of the mixed falts, I attempted to feparate the fulphate, fulphite, and carbonate of potafh, from the combination of nitrous oxide and potafh, by fucceffive evaporations and cryftalifations. But though in this way it was nearly freed from fulphate of potafh, yet the extreme and nearly equal folubility of the other falts, prevented me from completely feparating them from each other.

· By expofing, however, very finely pulverifed fulphite of potafh, mingled with alkali, for a great length of time to nitrous gas, it was almoft wholly converted into sulphate; and after the separation of this by solution, evaporation, and cryftalifation, at a low temperature, I obtained the new combination, mingled with very little carbonate of potafh, and ftill lefs of fulphite.

The minute quantity of fulphite chiefly appeared in very fmall cryftals; diftinct from the mafs of falt, which poffeffed no regular cryftalifation, and was almoft wholly compofed of the new compound, intimately mingled with a little carbonate. The new compound, as nearly as as I could eftimate from the quantity of nitrous oxide abforbed, confifted of about 3 alkali, to 1 of nitrous oxide, by weight.

It exhibited the following properties:

1. Its tafte was cauftic, and poffeffed of a pungency different from either potafh or carbonate of potafh.

2. It rendered vegetable blues green, which might poffibly depend upon the carbonate of potafh mixed with it.

3. Pulverifed charcoal mingled with a few grains of it, and inflamed, burnt with flight

ſcintillations. Projected into zinc in a ſtate of fuſion, a ſlight inflammation was produced.

4. When either ſulphuric, muriatic, or nitric acid was introduced to it under mercury, it gave out nitrous oxide, mingled with a little carbonic acid.

5. Thrown into a ſolution of ſulphurated hydrogene, gas was diſengaged from it, but in quantities too minute to be examined.

6. When carbonic acid was thrown into a ſolution of it in water, gas was diſengaged; on examination it proved to be nitrous oxide.

7. A concentrated ſolution of it kept in ebullition in a cylinder, confined by mercury, gave out a few globules of gas, which were too minute to be examined, and probably conſiſted of common air previouſly contained in the water.

c. In the experiments made to aſcertain theſe properties all the ſalt was expended, otherwiſe I ſhould have endeavoured to aſcertain what quantity of gas would have been liberated by heat from a given weight; and likewiſe what would have

been the effects of admixture of it with oil.
When fome of the mixed falt was mingled with
oil of turpentine, part of it was diffolved, and
the fluid became white ; but no gas was given
out. On this coarfe experiment, however, I
cannot place much dependance. If the combination of nitrous oxide and potafh is capable
of combining with oil without decompofition,
barytes and ftrontian* will probably feparate the
oil from it, and thus it may poffibly be obtained
in a ftate of purity.

In a rough experiment made on the converfion of nitrous gas into nitrous oxide, by concentrated folution of fulphite of potafh with
excefs of alkali, very little of the nitrous oxide
was abforbed. Hence it is probable that water
leffens the affinity of potafh for nafcent nitrous
oxide.

---

* Unlefs the fum of affinity of the potafh, oil, nitrous
oxide, and earths, fhould be fuch as to enable the nitrous
oxide to combine with the earth, whilft the oil and alkali
remained in combination, &c.

VIII. *Combination of Nitrous Oxide with Soda.*

The union of nitrous oxide with foda is effected in the fame manner as with potafh. The alkali, mingled by folution and evaporation, with either fulphite of foda, or of potafh, is expofed to nitrous gas; the nitrous oxide is condenfed by it at the moment of generation, and the combination effected.

As far as I have been able to obferve, nitrous oxide is not abforbed to fo great an extent by foda, as potafh.

I have not yet been able to obtain the combination of nitrous oxide with foda in its pure ftate. To the attainment of this end, difficulties identical with thofe noticed in the laft fection prefent themfelves. It is extremely difficult to procure the foda perfectly free from carbonic acid, and though by ufing fulphite of potafh the fulphate formed is eafily feparated, yet ftill evaporation and cryftalifation will not difengage the fulphite and carbonate from the new compound.

The compound of foda and nitrous oxide, mingled with a little fulphite and carbonate of foda, was rapidly foluble, both in warm and cold water, without effervefcence. Its folution, heated to ebullition, gave out no gas. The tafte of the folid falt was cauftic, and more acrid than that of the mixture of carbonate and fulphite of foda. When caft upon zinc in fufion, it burnt with a white flame. When heated to 400° or 500°, it gave out nitrous oxide with rapidity. Nitrous oxide was expelled from it by the fulphuric, muriatic, and carbonic acids, *I believe*, by fulphurated hydrogene.*

IX. *Combination of Nitrous Oxide with Ammoniac.*

I attempted to effect this combination by

---

* For when a little of the mixed falt was introduced into a folution of fulphurated hydrogene, globules of gas were given out during the folution.

converting nitrous gas into nitrous oxide, by sulphite of ammoniac, wetted with strong solu-tion of cauſtic ammoniac ; but, without fucceſs ; for the whole of the nitrous oxide produced remained in a free ſtate.

When I, expoſed ſulphite of potaſh, mingled by folution and evaporation with highly alkaline carbonate of ammoniac,† to nitrous gas, the diminution was nearly one fourth more than if pure ſulphite of potaſh had been employed. Hence it appears moſt likely that ammoniac is capable of combination with nitrous oxide in the naſcent ſtate.

In the experiments on the converſion of nitrous gas into nitrous oxide, by naſcent hydro-gene, and by ſulphurated hydrogene, Ref. I. Diviſ. V. probably the water formed at the fame

---

† Carbonate of ammoniac formed at a high tempera-ture, containing near 60 per cent alkali, and capable of combining with ſmall quantities of acids without giving out its carbonic acid. Of this ſalt a particular account will be given in the experiments on the ammoniacal ſalts, which I have often mentioned in the courſe of this work.

time with the ammoniac and nitrous oxide, prevented them from entering into combination;
poſſibly the peculiar compound was formed, but
in quantities ſo minute as not to be diſtinguiſhed
from ſimple ammoniac;* for even the exiſtence
of ammoniac in theſe proceſſes, is but barely
perceptible.

If it ſhould be proved by future experiments,
that in the decompoſition of nitrous gas by
naſcent hydrogene, a peculiar compound of
nitrous oxide, water and ammoniac, is formed,
it will afford proofs in favor of the doctrine
of prediſpoſing affinity;† for then this decom-

---

* It may not be amiſs to mention ſome appearances taking
place in the decompoſition of nitrous gas by ſulphurated
hydrogene, though it is uſeleſs to theoriſe concerning them.
The ſulphur depoſited is at firſt yellow; as the proceſs proceeds, it becomes white, and in ſome inſtances I have ſuſpected a diminution of it.

---

†.Predifpoſing affinity, the exiſtence of which at firſt
conſideration it is difficult to admit, may be eaſily accounted
for by *ſuppoſing* the attractions of the ſimple principles of
compound ſubſtances. And this doctrine will apply in all

pofition might be fuppofed to depend upon the difpofition of oxygene, hydrogene and nitrogene to affume the ftates of combination in which they might form a triple compound, of water, nitrous oxide, and ammoniac.

Nitrous oxide might probably be made to combine with ammoniac by expofing a mixture of nitrous gas and aëriform ammoniac, to the fulphites.

It is probable that nitrous oxide may be combined with ammoniac, by means of double affinity. Perhaps fulphate of ammoniac and the combination of potafh with nitrous oxide mingled together in folution, would be converted into fulphate of potafh and the compound of nitrous oxide, and ammoniac.

instances where the conftitution of bodies is known. Predifpofing affinity ought not to be confidered as the affinity of *non-exifting* bodies for each other; but as the mutual affinity of their fimple principles, difpofing them to affume new arrangements.

X. *Probability of forming Compounds of Nitrous Oxide and the Alkaline Earths.*

I attempted to combine nitrous oxide with lime and ftrontian, by expofing fulphites of lime and ftrontian with excefs of earth, to nitrous gas; but this procefs did not fucceed: the diminution took place fo flowly as to deftroy all hopes of gaining any refults in a definite time. Sulphite of potafh is decompofable by barytes, ftrontian, and lime;* confequently it was impoffible to employ this fubftance to effect the combination.

As the dry fulphures, when well made, convert nitrous gas into nitrous oxide, it is probable that the union of the earths with nafcent nitrous oxide may be effected by expofing nitrous gas to their fulphures, containing an excefs of earth.

Perhaps the combination of nitrous oxide with

---

* See the above-mentioned elaborate memoir of Fourcroy and Vauquelin.

ſtrontian may be effected by introducing the combination of potaſh and nitrous oxide into ſtrontian lime water.

It is probable that nitrous oxide may be combined with clay and magneſia, by expoſing theſe bodies, mingled with ſulphite of potaſh or ſoda, to nitrous gas.

XI. *Additional Obſervations on the combinations of Nitrous Oxide with the Alkalies.*

A deſire to complete phyſiological inveſtigations relating to nitrous oxide, has hitherto prevented me from purſuing to a greater extent, the experiments on the combination of this ſubſtance with the alkalies, &c. As ſoon as an opportunity occurs, I purpoſe to reſume the ſubject.

The obſervations detailed in the foregoing ſections are ſufficient to ſhow that nitrous oxide is capable of entering into intimate union with the fixed alkalies : and as the compounds formed by this union are inſoluble in alcohol,

decompofable by the acids, and heat, and poffeffed of peculiar properties, they ought to be
confidered as a new clafs of faline fubftances.

If it is thought proper, on a farther inveftigation of their properties, to fignify them by
fpecific names, they may, according to the ufually adopted fafhion of nomenclature, be called
nitroxis : thus the *nitroxi of potafh* would fignify
the falt formed by the combination of nitrous
oxide with potafh.

Future experiments muft determine the
different affinities of nitrous oxide for the alkalies, and alkaline earths.

With regard to the ufes of thefe new compounds it is difficult to form a guefs. When
they are obtained pure, and fully faturated with
nitrous oxide, on account of the low temperature
at which their gas is liberated, they will probably conftitute detonating compounds. From
their facility of decompofition by the weaker
acids, they may poffibly be ufed medicinally, if
ever the evolution of nitrous oxide in the ftomach
fhould be found beneficial in difeafes.

XII. *The properties of Nitrous Oxide resemble those of Acids.*

If we were inclined to generalise, and to place nitrous oxide among a known class of bodies, its properties would certainly induce us to consider it as more analogous to the acids than to any other substances; for it is capable of uniting with water and the alkalies, and is insoluble in most of the acids. It differs, however, from the stronger acids, in not possessing the four taste,* and the power of reddening vegetable blues: and from both the stronger and weaker acids, in not being combinable when in a perfectly free state, at common tempera-

---

* The different persons who have respired nitrous oxide have, as will be seen hereafter, given different accounts of the taste; the greater number have called it sweet, some metallic. One of my friends, in a letter to me dated Nov. 13, 1799, containing a detail of some experiments made on the respiration of nitrous oxide, at Birmingham, denotes the taste of it by the term " sweetish faintly acidulous." To me the taste both of the gas and of its solution in water, has always appeared faintly sweetish.

tures, with the alkalies. If it fhould be proved by future experiments, that condenfation by cold gave it the capability of immediately forming neutro-faline compounds with the alkalies; it ought to be confidered as the weakeft of the acids. Till thofe experiments are made, its extraordinary chemical and phyfiological properties are fufficient to induce us to confider it as a body *fui generis.*

It is a fingular fact that nitrous gas, which contains in its compofition a quantity of oxygene fo much greater than nitrous oxide, fhould neverthelefs poffefs no acid properties. It is uncombinable with alkalies, very little foluble in water, and abforbable by the acids.

DIVISION II.

On the DECOMPOSITION of NITROUS OXIDE by COMBUSTIBLE BODIES. Its ANALYSIS. OBSERVATIONS on the different combinations of OXYGENE and NITROGENE.

I. Preliminaries.

From the phænomena mentioned in Ref. I. Divif. III.* it appears that the combuftible bodies burn in nitrous oxide at certain temperatures. The experiments in this Divifion were inftituted for the purpofe of inveftigating the precife nature of thefe combuftions, with a view of afcertaining exactly the compofition of nitrous oxide.

It will be feen hereafter that very high temperatures are required for the decompofition of

---

* Section 2.

nitrous oxide, by moſt of the combuſtible bodies, and that in this proceſs heat and light are produced to a very great extent. Theſe agents alone are poſſeſſed of a conſiderable power of action on nitrous oxide; of which it is neceſſary to give an account, that we may be able to underſtand the phænomena in the following ſections.

II. *Converſion of Nitrous Oxide into Nitrous Acid, and a Gas analogous to Atmoſpheric Air, by Ignition.*

a. Dr. Prieſtley aſſerts, that nitrous oxide expoſed for a certain time to the action of the electric ſpark, is rendered immiſcible with water, and capable of diminution with nitrous gas, without ſuffering any alteration of volume; and likewiſe that the ſame changes are effected in it by expoſure to ignited incombuſtible bodies.*

The Dutch chemiſts ſtate, that the electric

* Vol. ii. pag. 91.

ſpark paſſed through nitrous oxide, occaſions a
ſmall diminution of its volume, and that the
gas remaining is analogous to common air.†
They conclude that this change depends on the
ſeparation of its conſtituent parts, oxygene and
nitrogene, from each other.

None of theſe chemiſts have ſuſpected the
production of nitrous acid in this proceſs.

b. Nitrous oxide undergoes no change
whatever from the ſimple action of light. I
expoſed ſome of it, confined by mercury, for
many days to this agent, often paſſing through
it concentrated rays by means of a ſmall lens.
When examined it appeared, as well as I could
eſtimate, of the ſame degree of purity as at the
beginning of the experiment.

c. A temperature below that of ignition
effects no alteration in the conſtitution of
nitrous oxide. I paſſed nitrous oxide from a

† Journal de Phyſique, tom. xliii, part ii. pag. 330. They
effected the ſame change by paſſing it through a heated
tube. Dr. Prieſtley had publiſhed an account of ſimilar
experiments more than two years before.

retort containing decomposing nitrate of ammoniac, through a green glass tube, strongly heated in an air-furnace, but not suffered to undergo ignition. The gas, received in a water apparatus exhibited the same properties as the purest nitrous oxide; some of it absorbed by water, left a residuum of not quite one thirteenth.

d. The action of the electric spark for a long while continued, converts nitrous oxide into a gas analogous to atmospheric air, and nitrous acid.

I passed about 150 strong shocks from a small Leyden phial, through 7 ten grain measures of pure nitrous oxide. After this it filled a space rather less than six measures: the mercury was rendered white on the top, as if it had been acted on by nitric acid. Six measures of nitrous gas mingled with the residual gas of the experiment, over mercury covered by a little water, gave red fumes, and rapid diminution. In five minutes the volume of the gases nearly equalled ten. Thermometer in this experiment was 58°.

Electric fparks were paffed for an hour and half through 7 ten grain meafures of nitrous oxide over mercury covered with a little red cabbage Juice, previoufly faturated with nitrous oxide, and rendered green by an alkali. After the procefs the gas filled a fpace equal to rather more than fix meafures and half, and the juice was become of a pale red. The gas was introduced into a fmall tube filled with pure water, and agitated; no abforption was perceptible: 7 meafures of nitrous gas added to it gave red fumes, and after fix minutes a diminution to $9\frac{1}{4}$ nearly. $6\frac{1}{2}$ meafures of common air from the garden, with 7 of nitrous gas, gave exactly 9.

In this experiment it was evident that nitrous oxide was converted into a gas analogous to atmofpheric air, at the fame time that an acid was formed. There could be little doubt but that this was the nitrous acid. To afcertain it, however, with greater certainty, the electric fpark was paffed through 6 meafures of nitrous oxide, over a little folution of green fulphate of iron, confined by mercury. As the procefs went on, the color of the folution became rather darker. When the diminution was complete, a little pruffiate of potafh was added to the folution. A precipitate of pale blue pruffiate of potafh was produced.

c. Nitrous oxide was paffed from decompofing nitrate of ammoniac, through a porcelain tube well-glazed infide and outfide, ftrongly ignited in an air-furnace, and communicating with the water apparatus. The gas collected was rendered opaque by denfe red vapor. It appeared wholly unabforbable by water. After the precipitation of its vapor, a candle burnt in it with nearly the fame brilliancy as in atmofpheric air. 20 meafures of it that had been agitated in water immediately after its production, mingled with 40 meafures of nitrous gas, diminifhed to about 47.5 ; whereas 20 meafures that had remained unagitated for fome time after their generation, introduced to the fame quantity of nitrous gas, gave nearly 49. 20 meafures of atmofpheric air, with 40 of the fame nitrous gas, were condenfed to 46.

The water with which the gas had been in contact, was ſtrongly acid. A little of it poured into a ſolution of green ſulphate of iron, and then mingled with pruſſian alkali, produced a green precipitate. Hence the acid it contained was evidently nitrous.

That no ſource of error could have exiſted in this experiment from fiſſure in the tube, I proved, by ſending water through it whilſt ignited, after the proceſs, from the ſame retort in which the nitrate of ammoniac had been decompoſed; a few globules of air only were produced, not equal to one tenth of the volume of the water boiled, and which were doubtleſs previouſly contained in it.

I have repeated this experiment two or three times, with ſimilar reſults; whenever the air was agitated in water immediately after its production, it gave *almoſt* the ſame diminution with nitrous gas as common air; when, on the contrary, it has been ſuffered to remain for ſome time in contact with the phlogiſticated nitrous acid ſuſpended in it, the condenſation has been leſs with nitrous gas by five or fix hundred parts. Hence I am inclined to believe, that if it were poffible to condenfe all the nitrous acid formed, immediately after its generation, fo as to prevent it from abforbing oxygene from the permanent gas, this gas would be found identical with the air of the atmofphere.

The changes effected by fire on nitrous oxide are not analogous to thofe produced by it in other bodies; for the power of this agent feems generally *uniform*, either in wholly feparating the conftituent principles of bodies from each other, or in making them enter into more intimate union.*

It is a fingular phænomenon, that whilft it condenfes one part of the oxygene and nitrogene of nitrous oxide, in the form of nitrous acid;

---

* On the one hand, it decompofes ammoniac into hydrogene and nitrogene, whilft on the other, it converts free oxygene and nitrogene into nitrous acid. It likewife converts nitrous gas into nitrous acid and nitrogene. Till we are more accurately acquainted with the nature of heat, light, and electricity, we fhall probably be unable to explain thefe phænomena.

it fhould, caufe the remainder to expand, in the ftate of atmofpheric air. Does not this fact afford an inference in favor of the *chemical* compofition of atmofpheric air ?

III. *Decompofition of Nitrous Oxide by Hydrogene, at the temperature of Ignition.*

In the following experiments on the decompofition of nitrous oxide by hydrogene, the gafes were carefully generated in the mercurial apparatus, and their purity afcertained by the tefts mentioned in Refearch I. They were meafured in fmall tubes graduated to grains, and then transferred into the detonating tube, which, was eight tenths of an inch in diameter, and graduated to ten grain meafures.

The fpace occupied by the gafes being noted after the inflammation by the electric fhock, green muriate of iron, and pruffiate of potafh, were fucceffively introduced, to afcertain if any nitrous acid had been formed. The abforption, if any took place, was marked, and the gafes transferred into a narrow grain measure tube, and their bulk and composition accurately ascertained.

b. The hydrogene employed was procured from water by means of zinc and sulphuric acid. · 50 grain measures of it fired by the electric spark, with 30 grain measures of oxygene: containing one eleventh nitrogene, gave a residuum of about 4. Nitrous gas mingled with those 4, indicated the presence of rather less than 1 of unconsumed oxygene. In another experiment 23 of it, with 20 of the same oxygene left rather more than 6 residuum. The nitrous oxide was apparently pure, for it left a remainder of about ,05 only, when absorbed by common water.

c. 50 of hydrogene were fired with 40. of nitrous oxide; the concussion was very great, and the light given out bright red; no perceptible quantity of nitrous acid was formed; the residual gas filled a space equal to 52. No part of it was absorbable by water, it gave no diminution with nitrous gas, when it was mingled with a little oxygene; and again acted on by the

electric fpark, an inflammation and flight diminution was produced.

d. 33 of hydrogene were fired with 35 of
nitrous oxide : nitrous acid was produced in
very minute quantity ; the gas that remained
was not abforbable by water, and filled a fpace
equal to 37 grains. Nitrous gas mingled with
thefe, underwent a very flight diminution.

e. 46 hydrogene were fired with 46 nitrous
oxide. The quantity of nitrous acid formed
was juft fufficient to tinge the white pruffiate
of potafh. The gafes filled a fpace equal to 49,
gave no perceptible diminution with nitrous
gas, and did not inflame with oxygene.

f. 40 hydrogene were fired with 39 nitrous
oxide ; no perceptible quantity of nitrous acid
was formed. The refidual gas filled a fpace
equal to 41 ; was unabforbable by water, underwent no diminution when mingled with
nitrous gas ; or when acted on by the electric
fpark in contact with oxygene.

g. 20 hydrogene were fired with 64 nitrous
oxide ; after detonation the expanfion of the

gafes was greater in this experiment than
any of the preceding ones ; denfe white
fumes were obferved in the cylinder, and
a flow contraction of volume took place.
After a little green muriate of iron had been
admitted, the gafes filled a fpace equal to
73 : pruffiate of potafh mingled with the muriate, gave a deeper blue than in any of the
preceding experiments. The refidual gas was
unabforbable by water : .65 of it, mingled with
65 of nitrous gas, diminifhed to 93 ; whilft 65
of common air, with 65 of nitrous gas, gave
84.

h. 18 of hydrogene were fired with 54 of
nitrous oxide ; the fame phænomena as were
obferved in the laft experiment took place ;
nitrous acid was formed ; after the abforption of
which the refidual gas filled a fpace equal to
55. 50 of this, with an equal quantity of
nitrous gas, diminifhed to 76. In thefe proceffes the temperatures were from 56° to 61°.

These experiments are felected as the moft
accurate of nearly fifty, made on the inflamma-

tion of different quantities of nitrous oxide and hydrogene.

As Mr. Keir found muriatic acid in the fluid, produced by the inflammation of oxygene and hydrogene in clofed veffels, in Dr. Prieftley's experiments, I preferved the refidual gas of about 3 cubic inches of nitrous oxide, that had been detonated at different times with lefs than a cubic inch and half of hydrogene; but folution of nitrate of filver was not clouded when agitated in this gas, nor when introduced into the detonating tube in which the inflammation had been made.

From thefe experiments we learn that nitrous oxide is decompofable at the heat of ignition, by hydrogene, in a variety of proportions.

When the quantity of hydrogene very little exceeds that of the nitrous oxide, both of the gafes difappear, water is produced, no nitrous acid is formed, and the volume of nitrogene evolved is rather greater than that of the nitrous oxide decompofed.

When the quantity of hydrogene is lefs than

that of the nitrous oxide, water, nitrous acid, oxygene, and nitrogene, are generated in different proportions; one part of the nitrous oxide is moſt probably wholly decompoſed by the hydrogene, and the other part converted into nitrous acid and atmoſpheric air, in conſequence of the ignition.

From experiments *c*, *d*, and *e*, the compoſition of nitrous oxide may be deduced. In experiment *d*, .39 of nitrous oxide were decompoſed by 40 of hydrogene, and converted into 41 of nitrogene.

Now from *b* it appears that 40 of hydrogene require for their condenſation about 20.8 of oxygene in volume; ſo that founding the eſtimation upon the quantity of hydrogene conſumed, 100 parts of nitrous oxide would conſiſt nearly of 63.1 of nitrogene, and 36.9 of oxygene. But 41 of nitrogene weigh 12.4, Ref. I. Div. I. Conſequently, deducing the compoſition of nitrous oxide from the quantity of nitrogene evolved, 100 parts of it would conſiſt of 63.5 nitrogene, and 36.5 oxygene.

These estimations are very little different from those which may be deduced from the other experiments, and the coincidence is in favor of their accuracy.

From the following experiment it appears that the temperature required for the decomposition of nitrous oxide by hydrogene must be higher than that which is necessary to produce the inflammation of hydrogene with oxygene. I introduced into small tubes filled with equal parts of nitrous oxide and hydrogene, standing on a surface of mercury, iron wires ignited to different degrees, from the dull red to the vivid white heat. The gases were always inflamed by the white and vivid red heats; but never by the dull red heat, though the last uniformly inflamed mixtures of oxygene and hydrogene, and atmospheric air and hydrogene.

Dr. Priestley * first detonated together nitrous oxide and hydrogene; his experiment was repeated by the Dutch chemists, who found that when a small quantity of hydrogene was

* Vol. ii, pag. 83.

employed, the nitrous oxide was partially converted into a gas analogous to common air.
Their eftimation of its compofition, which is
not far removed from the truth, was founded
on this phænomenon.*

IV. *Decompofition of Nitrous Oxide by Phofphorus.*

' a. . Phofphorus introduced into pure nitrous
oxide at common temperatures, is not at all
luminous. It is capable of being fufed, and
even fublimed in it, without undergoing acidifi-
cation, and without effecting any alteration in
its compofition.

About 2 grains of phofphorus were fufed, and
gradually fublimed, in 2 cubic inches of pure
nitrous oxide, over mercury, by the heat of a

* Journal de Phyfique, tom. xliii. part 2, pag. 331.
They fuppofed it to confift of about 37,5 oxygene, and
62,5 nitrogene. The nearnefs of this account to the truth
is fingular, when we confider that they were neither acquainted with the fpecific gravity of nitrous oxide, nor
with the production of nitrous acid in this experiment.

burning lens. No alteration was produced in the volume of gas, and a portion of it abforbed by water, left a refiduum of one twelfth only.

Phofphorus was fublimed in pure nitrous oxide over mercury, in a dark room, by an iron heated *nearly* to ignition ; but no luminous appearance was perceptible, nor was any gas decompofed.

b. Phofphorus decompofes nitrous oxide at the temperature of ignition, with greater or lefs rapidity, according to the degree of heat. We have already feen, that when phofphorus in active inflammation is introduced into nitrous oxide, it burns with intenfely vivid light.

Phofphorus was fublimed by a heated wire in a jar filled with nitrous oxide, ftanding over warm mercury. In this ftate of fublimation an iron heated dull red was introduced to it by being rapidly paffed through the mercury; a light blue flame furrounded the wire, and difappeared as foon as it ceafed to be red.

To phofphorus fublimed as before, in nitrous oxide, over warm mercury, a thick wire ignited

to whitenefs was introduced ; a terrible detonation took place, and the jar was fhattered in
pieces.

By employing thick conical jars,* containing
only a fmall quantity of nitrous oxide, I effected
the detonation feveral times with fafety ; but
on account of the great expanfion of the elaftic
products, the jar was generally either raifed
from the mercury, or portions of gas were
thrown out of it. Hence I was unable to afcertain the exact changes produced by this mode
of decompofition.

c. As my firft attempts to afcertain the conftitution of nitrous oxide were made on its decompofition by phofphorus, I employed many dif-

---

* Experiments on the detonatiou of nitrous oxide with
phofphorus in this way require great attention. The detonating jar fhould be very conical; the nitrous oxide employed fhould never equal more than one eighth of the
capacity of the jar. The wire for the inflammation muft
be very thick, and curved fo as to be eafily introduced into
the jar. When ignited, it muft be inftantaneoufly paffed
through the heated mercury into the jar.

Perhaps the electric fpark might be advantageoufly applied for detonating phofphoric vapor with nitrous oxide.

ferent modes of partially igniting this fubftance
in it over mercury, fo as to produce a combuftion without explofion.

The firft method adopted was inflammation
by means of oxygenated muriate of potafh. A
fmall particle of oxygenated muriate of potafh
was inferted into the phofphorus to be burnt.
On the application of a wire, moderately hot,
to the point of infertion, the falt was decompofed by the phofphorus, and fufficient fire
generated and partially applied by the flight
explofion, to produce the combuftion of the
phofphorus, without the previous fublimation
of any part of it.

The fecond way employed was the ignition
of a part of the phofphorus, by means of the
combuftion of a fmall portion of tinder of cot-
ton,* or paper, in contact with it, by the
burning glafs.

The third, and moft fuccefsful mode, was
by introducing into the graduated jar containing

---

* It will be feen hereafter that thefe bodies are eafily
inflamed in nitrous oxide.

the nitrous oxide, the phofphorus in a fmall
tube containing oxygene, fo balanced as to
fwim on the furface of the mercury, without
communicating with the nitrous oxide. The
phofphorus was fired in the oxygene with an
ignited iron wire, by which at the moment of
combuftion, the tube containing it was raifed
into the nitrous oxide, and thus the inflammation continued.

d. In different experiments, made with accuracy, I found that the whole of a quantity of nitrous
oxide was never decompofable by ignited phofphorus; the combuftion always ftopped when
the nitrous oxide remaining was to the nitrogene evolved as about 1 to 5 ; likewife that the
volume of nitrogene produced was rather lefs
than that of the nitrous oxide decompofed, and
that this deficiency arofe from the formation of
nitrous acid by the intenfe ignition produced
during the procefs.

Of one experiment I fhall give a detail.

Temperature being 48°, two cubic inches
of pure nitrous oxide, which had been generated

over mercury, were introduced into a jar of the
capacity of 9 cubic inches, graduated to 1 cubic
inches, and much enlarged at the base. A grain
of phosphorus was inserted into a small vessel
about one third of an inch long, and half an
inch in diameter, containing about 15 grain
measures of very pure oxygene; this vessel,
which swam on the surface of the mercury, was
carefully introduced into the jar containing the
nitrous oxide. The phosphorus was fired by
means of a heated wire; and before the oxygene
was wholly consumed, the vessel containing it
elevated into the nitrous oxide. The combustion was extremely vivid and rapid. After
the atmospheric temperature was restored, the
gas was rendered opaque by dense white vapor.
When this had been precipitated, and the small
vessel removed from the jar, the gas filled a
space nearly equal to 1.9 cubic inches. On
introducing to it a little solution of green muriate of iron, and prussiate of potash, green
prussiate of iron was produced: hence, evidently, nitrous acid had been formed.

On the admiffion of pure water, an abforption of rather more than ,3, took place.

The 16 meafures remaining, underwent no perceptible diminution with nitrous gas ; the taper plunged,into them was inftantly extinguifhed.

To afcertain if the phofphoric acid produced in the experiments made under mercury did not in fome meafure prevent the decompofition of the whole of the nitrous oxide by the phofphorus, I introduced into a mixture of 5 nitrogene and 1 nitrous oxide, ignited phofphorus : but it was immediately extinguifhed.*

The Dutch Chemifts found that phofphorus might be fufed in nitrous oxide without being luminous. They affert that phofphorus in a ftate of inflammation, introduced into this gas, was immediately extinguifhed ; though when taken out into the atmofphere, it again burnt of its own accord.† It is difficult to account for their miftake.

* Phofphorus burnt feebly with a white flame in a mixture of 4 nitrogene and 1 nitrous oxide.

V. *Decompofition of Nitrous Oxide by Phof-phorated Hydrogene.*

a. It has been mentioned in Ref. II. Div. I. that phofphorated hydrogene and nitrous oxide poffefs no action on each other, at atmofpheric temperatures.

Phofphorated hydrogene mingled with nitrous oxide, is capable of being inflamed by the electric fpark, or by ignition.

b. E. 1. 10 grain meafures of phofphorated hydrogene, carefully produced by means of phofphorus and folution of cauftic alkali, were mingled with 52 meafures of nitrous oxide. The electric fpark paffed through them, produced a vivid inflammation. The elaftic products were clouded with denfe white vapor, and after fome minutes filled a fpace nearly equal to 60. On the introduction of water, no abforption took place. When 43 of nitrous gas were admitted, the whole diminifhed to 70.

E. 2. 25 of nitrous oxide were fired with 10

of phofphorated hydrogéne, by the electric
fpark. After, detonation*, they filled a fpace
exactly : equal to 25. On the admiffion of
folution of green fulphate of iron, and pruffiate
of potafh, no blue or green precipitate was produced. On the introduction of water, no diminution was perceived. - 25 of nitrous gas mingled with them, gave exactly 50.

E. 3. 10 of nitrous oxide, mingled with 20 of
phofphorated hydrogene, could not be inflamed.

25 of nitrous oxide, with 20 phofphorated
hydrogene, inflamed. The gas after detonation,
was rendered opaque by denfe white vapor, and
filled a fpace nearly equal to 45. No abforption
took place when water was introduced. On
admitting a little oxygene no white fumes, or
diminution, was perceived. The electric fpark
paffed through the mixture, produced an explofion, with great diminution.

c. From *E.* 1 it appears, that when a fmall quantity of phofphorated hydrogene is inflamed with

---

* In this experiment, as in the laft, denfe white vapor
was produced.

nitrous oxide, both the phofphorus and hydrogene are confumed ; whilft the fuperabundant
nitrous oxide, is converted into nitrous acid and
atmofpheric air, by the ignition ; or a certain
quantity is partially decompofed into atmofpheric
air by the combination of a portion of its oxygene
with the combuftible gas.

From *E*.2 we learn, that when the phofphorated
hydrogene and nitrous oxide are to each other
as 25 to 10 nearly, they both difappear, whilft
nitrogene is evolved, and water and phofphoric
acid produced. Reafoning concerning the
compofition of nitrous oxide from this experiment, we fhould conclude that it was compofed
of about 38 oxygene, and 62 nitrogene.

The refult of *E*. 3 is interefting; we are taught
from it that the affinity of phofphorus for the
oxygene of nitrous oxide is ftronger than that
of hydrogene, at the temperature of ignition ;
fo that when phofphorated hydrogene is mingled with a quantity of nitrous oxide, not containing fufficient oxygene to burn both its conftituent parts, the phofphorus only is confumed,
whilft the hydrogene is liberated.

,In repeating the experiments with phofphorated hydrogene that had remained, for fome
hours in the mercurial apparatus, I did not gain
exactly the fame refults; for a larger quantity
of it was required to decompofe the nitrous
oxide, than in the former experiments; doubtlefs from its having depofited a portion of its
phofphorus. They confirm, however, the
above mentioned conclufions.

In the courfe of experimenting, I paffed the
electric fpark, for a quarter of an hour, through
about 60 meafures of phofphorated hydrogene.
It underwent no alteration of volume. Phofphorus was apparently precipitated from it, and
it had wholly loft its power of inflaming, in
contact with common air.

VI. *Decompofition of Nitrous Oxide by Sulphur.*

From the phænomena before mentioned,[*]

relating to the combuftion of fulphur in nitrous oxide, it was evident that this gas was only de-compofable by it, at a much higher temperature than common air.

I introduced into fulphur, in contact with nitrous oxide, over mercury heated to 112°—114°, a wire intenfely ignited. It loft much of its heat in paffing through the mercury, but ftill appeared red in the veffel. The fulphur rapidly fufed, and fublimed without being at all luminous. This experiment was repeated five or fix times, but in no inftance could the combuftion of fulphur, by means of the ignited wire, be effected.

I inflamed fulphur in nitrous oxide in the fame manner as phofphorus; namely, by intro-ducing it into the fmall veffel filled with oxy-gene, and igniting it by means of the heated wire. In thefe experiments the fulphur burnt with a vivid rofe-colored light, and much ful-phuric, with a little fulphureous acid, was formed.

Experimenting in this way I was never, how-ever, able to decompofe more than one third

of the quantity of nitrous oxide employed;
not only the nitrogene evolved, but likewife the
fulphuric and fulphureous acids produced, flop-
ping the combuftion.

I found that fulphur in a ftate of vivid inflammation, when introduced into a mixture of
one fourth nitrogene, and three fourths nitrous
oxide, burnt with a flame very much enlarged,
and of a vivid rofe color. In one third nitrogene, and two thirds nitrous oxide, it burnt
feebly with a yellow flame. In equal parts of
nitrous oxide and nitrogene, it was inftantly
extinguifhed.

Sulphur burnt feebly, with a light yellow
flame, when introduced ignited into a mixture
of 5 nitrous gas, and 6 nitrous oxide. In one
third nitrous oxide, and two thirds nitrous gas,
it was inftantly extinguifhed. From many
circumftances, I am inclined to believe that fulphur is incapable, at any temperature, of flowly
decompofing nitrous oxide, fo as to burn in it
with a blue flame, forming fulphureous acid
alone. It appears to attract oxygene from it

U

only when intenſely ignited, ſo as to form chiefly ſulphuric acid, and that with great rapidity, and vivid inflammation.

VII. *Decompoſition of Nitrous Oxide by Sulphurated Hydrogene.*

a. Though nitrous oxide and ſulphurated hydrogene do not act upon each other at common temperatures, yet they undergo a mutual decompoſition when mingled together in certain proportions, and ignited by the electric ſpark.

From more than twenty experiments made on the inflammation of ſulphurated hydrogene in nitrous oxide, I ſelect the following as the moſt concluſive and accurate. The temperature at which they were made was from 41° to 49°.

b. *E.* 1. About 35 meaſures of nitrous oxide were fired with 10 of ſulphurated hydrogene; the expanſion during inflammation was very great, and the flame ſky-blue. Immediately after, the gaſes filled a ſpace equal to 48 nearly. White fumes were then formed, and they gradually contracted to 40. On the admiffion of a little ftrontian lime water, a flight abforption took place, with white precipitation ; and the volume occupied by the refidual gas nearly equalled 37. On admitting nitrous gas to thefe, no perceptible diminution took place.

E. 2. 20 fulphurated hydrogene, with 25 nitrous oxide, could not be inflamed.

30 nitrous oxide, with 22 fulphurated hydrogene, could not be inflamed.

35 nitrous oxide, with 20 fulphurated hydrogene, inflamed with vivid blue light, and great expanfion. After the explofion, the gafes filled exactly the fame fpace as before the experiment; no white fumes were perceived, and no farther contraction occurred. On the addition of ftrontian lime water, a copious precipitation, with diminution, took place ; and the refidual gas filled a fpace nearly equal to $35\frac{1}{2}$.

E. 3. 47 nitrous oxide, and 14 fulphurated hydrogene, inflamed. After the explofion, the gafes filled a fpace nearly equal to 65 ; then white fumes formed, and they gradually diminifhed to 52. On the introduction of muriate of ftron-

tian, a copious white precipitate was produced;
and on the addition of water, no further abforption took place. To the refidual 52, about
20 of nitrous gas were added; they filled together a fpace equal to about 67.

c. In none of the experiments made on the
inflammation of fulphurated hydrogene and
nitrous oxide, could I afcertain with certainty
the precipitation of fulphur. In one or two
proceffes the detonating tube was rendered a
little white at the points of contact with the
mercury; but this was moft probably owing to
the oxydation of the mercury, either by the
heated fulphuric acid formed, or from nitrous
acid produced by the ignition. The prefence
of nitrous acid I could not afcertain in thefe proceffes by my ufual tefts, becaufe the combuftion
of fulphur over white pruffiate of iron, converts
it into light green.

When I introduced an inflamed taper into
about 3 parts of fulphurated hydrogene, and 2
parts of nitrous oxide, in which proportions
they could not have been fired by the

electric fpark, a blue flame paffed through them, and much fulphur was depofited on the fides of the veffel. But this fulphur moft probably owed its formation to the decompofition of a portion of fulphurated hydrogene not burnt, by the fulphureous acid formed from the combuftion of the other portion.

We may then conclude with probability, that fulphurated hydrogene and nitrous oxide will not decompofe each other, when acted on by the electric fpark, unlefs their proportions are fuch as to enable the whole of the fulphurated hydrogene to be decompofed, fo that both of its conftituents may become oxygenated, by attracting oxygene from the nitrous oxide : likewife, that when the fulphurated hydrogene is at its *maximum* of inflammation, the hydrogene and fulphur form with the whole of the oxygene of nitrous oxide, water and fulphureous acid ; *E.* 2 : whereas at its *minimum* they produce water, and chiefly, *perhaps* wholly, fulphuric acid ; at the fame time that the nitrous oxide partially decompofed, is converted into nitrogene, and a

gas analogous to atmofpheric air, or into nitrogene, nitrous acid, and atmofpheric air. *E.* 1;
E. 3.

By purfuing thofe experiments, and ufing larger
quantities of gas, we may probably be able to
afcertain from them with accuracy, the compofition of fulphuric and fulphureous acids.

I own I was difappointed in the refults, for I
expected to have been able to afcertain from
them, the relative affinities of fulphur, and
hydrogene for the oxygene of nitrous oxide,
at the temperature of ignition. I conjectured
that nitrous oxide, mingled with excefs of fulphurated hydrogene, would have been decompofed, and one of the principles of it evolved
unaltered, as was the cafe with phofphorated
hydrogene.

If we eftimate the compofition of nitrous
oxide from the quantity of nitrogene produced in
E. 2, it is compofed of about 61 nitrogene, and
'39 oxygene.

An account of the analyfis of nitrous oxide by charcoal is given in Ref. I. Div. III. I have lately made two experiments on the combuftion of charcoal in nitrous oxide, in which every precaution was taken to prevent the exiftence of fources of error. Of one of thefe I fhall give a detail.

E. Temperature being 51°, about a grain of charcoal, which had been expofed for fome hours to a red heat, was introduced whilft ignited, under mercury, and transferred into a graduated jar, containing 3 cubic inches of pure nitrous oxide, ftanding over dry mercury.

The focus of a burning lens was thrown on the charcoal; it inftantly inflamed, and burnt with great vividnefs for near a minute, the gas being much expanded. The focus was continually applied to it for ten minutes, when the procefs appeared at an end. The gafes, when the common temperature and preffure were reftored, filled a fpace equal to 4,2 cubic inches.

On introducing into them a few grain mea-fures of folution of green muriate of iron, for the double purpofe of faturating them with moifture, and afcertaining if any nitrous acid had been formed, no change of volume took place; and pruffiate of potafh gave with the muriate a white precipitate only.

On the admiffion of a fmall quantity of con-centrated folution of cauftic potafh, a diminu-tion of the gas flowly took place ; when it was complete the volume equalled about 3.05 cubic inches. By agitation in well boiled water, about ,9 of thefe were abforbed ; the remainder appeared to be pure nitrogene.

The difference between the eftimation founded upon the nitrogene evolved, and that deduced from the carbonic acid generated in this experi-ment, is not nearly fo great as in that Ref. I. Div. III. Taking about the mean proportions, we fhould conclude that nitrous oxide was com-pofed of about 38 oxygene, and 62 nitrogene.

Charcoal burnt with greater vividnefs than in common air, in a mixture of one third nitrogene and two thirds nitrous oxide. In equal parts of nitrous oxide and nitrogene, its light was barely perceptible. In one third nitrous oxide, and two thirds nitrogene, it was almoft immediately extinguifhed.

As charcoal burns vividly in nitrous gas, when it has been previoufly ignited to whitenefs, I introduced it into a mixture of equal parts of nitrous oxide and nitrous gas ; it burnt with a deep and bright red.

IX. Decompofition of Nitrous Oxide by Hydro-carbonate.

Nitrous oxide, and hydro-carbonate, poffefs no action on each other, except at high temperatures. When mingled in certain proportions, and expofed to the electric fhock, a new arrangement of their principles takes place.

E. 1. Temperature being 53°, .35 of nitrous oxide, mingled with 15 of hydrocarbonate, were fired by the electric fpark; the inflammation was very vivid, and the light produced, bright red. After the explofion, the fpace occupied by the gafes equalled about 60. On the admiffion of folution of ftrontian, a copious white precipitate was produced, and the gas diminifhed by agitation, to rather more than 35. When 36 of nitrous gas were added to thefe, white fumes appeared and the whole diminifhed to 62. When a little muriatic acid was poured on the white precipitate from the folution of ftrontian, gas was evolved from it, and it was gradually diffolved.

E. 2. 22 nitrous oxide were inflamed with 20 hydro-carbonate; after the explofion, they filled a fpace equal to 45; when ftrontian lime water was introduced, white precipitation took place, and the diminution was to 31.

To thefe 31, 14 of nitrous oxide were admitted, and the electric fpark paffed through them; an inflammation took place: carbonic acid was formed, after the abforption of which, the gas remaining filled a fpace equal to 43, and did not diminifh with nitrous gas.

The hydro-carbonate employed in thefe experiments, was procured from alcohol by means of fulphuric acid. In another fet of experiments made with lefs accuracy, the fame general refults were obtained. Whenever hydro-carbonate inflamed with nitrous oxide, both its conftituents were oxygenated ; in all cafes carbonic acid was formed, and in no inftance free hydrogene evolved, or charcoal precipitated.

In the decompofition of nitrous oxide by hydro-carbonate, the refidual nitrogene is lefs than in in other combuftions. This circumftance I am unable to explain.

Reafoning from analogy, there can be little doubt, but that when hydro-carbonate is inflamed with excefs of nitrous oxide, it will be only partially decompounded, or converted into nitrogene, nitrous acid, and atmofpheric air.

The Dutch Chemifts have afferted, that charcoal does not burn in nitrous oxide, except in confequence of the previous decompofition of the gas by the hydrogene always contained in this fubftance; and likewife, that when hydro-carbonate and nitrous oxide were mingled together, and fired by the electric fpark, the hydrogene only was burnt, whilft the charcoal was precipitated.

It is difficult to account for thefe numerous miftakes. Their theory of the *non-refpirability* of nitrous oxide was founded upon them. They fuppofed that the chief ufe of refpiration was to deprive the blood of its fuperabundant carbon, by the combination of atmofpheric oxygene with that principle; and that nitrous oxide was highly fatal to life, becaufe it was incapable of de-carbonating the blood* ! !

X. *Combuftion of Iron in Nitrous Oxide.*

I introduced into a jar of the capacity of 20 cubic inches, containing 11 cubic inches of nitrous oxide, over mercury, a fmall quantity of fine iron wire twifted together, and having

* Journal de Phyfique, xliii. 334.

affixed to it a particle of cork. On throwing
the focus of a burning glafs on the cork, it
inftantly inflamed, and the fire was communicated to the wire, which burnt with great
vividnefs for fome moments, projecting bril-
liant white fparks. After it had ceafed to burn
the gas was increafed in volume rather more
than three tenths of an inch. The nitrous acid
tefts were introduced, but no acid appeared to
have been formed. On expofing the gas to
water, near 4,2 cubic inches were abforbed: the
7,1 remaining appeared to be pure nitrogene.

From this experiment it is evident that iron
at the temperature of ignition, is capable of
decompofing nitrous oxide; likewife that it is
incapable of burning in it when it contains more
than three fifths nitrogene.

I attempted to inflame zinc in nitrous oxide,
in the fame way as iron; but without fuccefs.
By keeping the focus of a burning glafs upon
fome zinc filings, in a fmall quantity of nitrous
oxide, I converted a little of the zinc into white
oxide, and confequently decompofed a portion
of the gas.

XI. *Combustion of Pyrophorus in Nitrous Oxide.*

Pyrophorus, which inflames in nitrous gas, and atmospheric air, at or even below 40°, requires for its combustion in nitrous oxide a much higher temperature. It will not burn in it, or alter it, even at 212°.

I have often inflamed pyrophorus in nitrous oxide over mercury, by means of a wire strongly heated, but not ignited. The light produced by the ignition of pyrophorus in nitrous oxide is white, like that produced by it in oxygene : in nitrous gas it is red.

When pyrophorus burns out in nitrous oxide, a little increase of the volume of gas is produced. Strontian lime water agitated in this gas becomes clouded; but the quantity of carbonic acid formed is extremely minute. I have never made any delicate experiments on the combustion of pyrophorus in nitrous oxide.

XII. *Combuſtion of the Taper in Nitrous Oxide.*

It has been noticed by different experi-mentaliſts, that the taper burns with a flame conſiderably enlarged in nitrous oxide: ſome-times with a vivid light and crackling noiſe, as in oxygene; at other times with a white central flame, ſurrounded by a feeble blue one.

My experiments on the combuſtion of the taper in nitrous oxide, were chiefly made with a view to aſcertain the cauſe of the double flame.

When the inflamed taper is introduced into pure nitrous oxide, it burns at firſt with a bril-liant white light, and ſparkles as in oxygene. As the combuſtion goes on, the brilliancy of the flame diminiſhes; it gradually lengthens, and becomes ſurrounded with a pale blue cone of light, from the apex of which much unburnt charcoal is thrown off, in the form of ſmoke. The flame continues double to the end of the proceſs.

When the refidual gafes are examined after combuftion, much nitrous acid is found fuf-pended in them; and they are compofed of carbonic acid, nitrogene, and about one fourth of undecompounded nitrous oxide.

The double flame depends upon the nitrous acid formed by the ignition; for it can be produced by plunging the taper into common air containing nitrous acid vapor, or into a mixture of nitrous oxide and nitrogene, through which nitrous acid has been diffufed. It is never perceived in the combuftion of the taper, till much nitrous acid is formed.

In attempting to refpire fome refidual gas of nitrous oxide, in which a taper had burnt out, I found it fo highly impregnated with nitrous acid, as to difable me from even taking it into my mouth.

The taper burns in a mixture of equal parts nitrous oxide and nitrogene, at firft with a flame nearly the fame as that of a candle in common air; white. Before its extinction the interior white flame, and exterior blue flame, are perceived.

The taper is inftantly extinguifhed in a mixture of one fourth nitrous oxide, and three
fourths nitrogene.

In a mixture of equal parts nitrous oxide and
nitrous gas, the taper burns at firft with nearly
as much brilliancy as in pure nitrous oxide;
gradually the double and feeble flame is produced.

XIII. *On the Combuftion of different Compound Bodies in Nitrous Oxide.*

All the folid and fluid compound inflammable
bodies on which I have experimented, burn in
nitrous oxide, at high temperatures. Wood, cot-
ton, and paper, are eafily inflamed in it by the
burning glafs. During their combuftion, nitrous acid is always formed, carbonic acid, and
water produced, and nitrogene evolved, rather
lefs in bulk than the nitrous oxide decompofed.
--I have already mentioned that alcohol and
ether are foluble in nitrous oxide. When an
ignited body is introduced into the folution of

W

alcohol, or ether in nitrous oxide, a flight explosion takes place.

XIV. *General Conclusions relating to the Decomposition of Nitrous Oxide, and to its Analysis.*

From what has been said in the preceding sections, it appears that the inflammable bodies, in general, require for their combustion in nitrous oxide, much higher temperatures than those at which they burn in atmospheric air, or oxygene.

When intensely heated they decompose it, with the production of much heat and light, and become oxygenated.

During the combustion of solid or fluid bodies, producing flame, in nitrous oxide, nitrous acid is generated, most probably from a new arrangement of principles, analogous to those observed in Sect. II, by the ignition of that part of the gas not in contact with the burning substance. Likewise when nitrous oxide in excess is decom-

pofed by inflammable gafes, nitrous acid, and
fometimes a gas analogous to common ·air, is
produced, doubtlefs from the fame caufe.

Pyrophorus is the only body that inflames in
nitrous oxide, below the temperature of
ignition.

Phofphorus. burns. in it with the blue flame,
probably forming with its oxygene only, phof-
phoreous acid at the dull red heat, and with the
intenfely vivid flame, producing phofphoric acid.
at the white heat.

Hydrogene, charcoal, fulphur, iron, and the
compound inflammable bodies, decompofe it
only at heats equal to, or above, that of ignition :
probably each a different temperature.

From the phænomena in Sect. V, it appears,
that at the temperature of intenfe ignition, phofphorus has a ftronger affinity for the oxygene
of nitrous oxide than hydrogene; and reafoning
from the different degrees of combuftibility of
the inflammable bodies, in mixtures of nitrous
oxide and nitrogene, and from other phæno-

mena, we may conclude with probability, that at about the white heat, the affinity of the combustible bodies for oxygene takes place in the following order. Phosphorus, hydrogene, charcoal,* iron, sulphur, &c.

This order of attraction is very different from that obtaining at the red heat; in which temperature charcoal and iron have a much stronger affinity for oxygene than either phosphorus or hydrogene.†

The smallest quantity of oxygene given in the different analyses of nitrous oxide just detailed, is thirty five hundred parts; the greatest proportion is thirty-nine.

Taking the mean estimations from the most accurate experiments, we may conclude that 100 grains of the known ponderable matter of

---

* As is proved by the decomposition of oxide of iron and sulphuric acid by charcoal, at that temperature.

† Hydrogene at or about the red heat, appears to attract oxygene stronger than phosphorus. See Dr. Priestley's experiments, vol. i. page 262.

nitrous oxide, confift of about 36,7 oxygene,
and 63,3 nitrogene; or taking away decimals;
of 37 oxygene to 63 nitrogene; which is identical
with the eftimation given in Refearch I.

XV. *Obfervations on the combinations of
Oxygene and Nitrogene.*

During the decompofitions of the combinations of oxygene and nitrogene by combuftible
bodies, a confiderable momentary expanfion of
the acting fubftances, and the bodies in contact
with them is generally produced, connected with
increafed temperature; whilft light is often
generated to a great extent.

Of the caufes of thefe phænomena we are at
prefent ignorant. Our knowledge of them
muft depend upon the difcovery of the precife
nature of heat and light, and of the laws by
which they are governed. The application of
general hypothefes to ifolated facts can be of
little utility; for this reafon I fhall at prefent
forbear to enter into any difcuffions concerning

thofe agents, which are imperceptible to the fenfes, and known only by folitary effects.

Analyfis and fynthefis clearly prove that oxygene and nitrogene conftitute the known ponderable matter of atmofpheric air, nitrous oxide, nitrous gas, and nitric acid.

That the oxygene and nitrogene of atmof-pheric air exift in chemical union, appears almoft demonftrable from the following evi-dences.

1ft. The equable diffufion of oxygene and nitrogene through every part of the atmofphere, which can hardly be fuppofed to depend on any other caufe than an affinity between thefe principles.*

2dly. The difference between the fpecific

---

* That attraction muft be called chemical, which enables bodies of different fpecific gavities to unite in fuch a manner as to produce a compound, in every part of which the con-ftituents are found in the fame proportions to each other. Atmofpheric air, examined after having been at perfect reft in clofed veffels, for a great length of time, contains in every part the fame proportions of oxygene and nitrogene; whereas if no affinity exifted between thefe principles, following the laws of fpecific gravity, they ought to fepa-

gravity of atmofpheric air, and a mixture of 27 parts oxygene and 73 nitrogene, as found by calculation ; a difference apparently owing to expanfion in confequence of combination.

3dly. The converfion of nitrous oxide into nitrous acid, and a gas analogous to common air, by ignition.

4thly. The folubility of atmofpheric air un-decompounded in water.

ATMOSPHERIC AIR, then, may be confidered as the leaft intimate of the combinations of nitrogene and oxygene.

It is an elaftic fluid, permanent at all known temperatures, confifting of ,73 nitrogene, and ,27 oxygene. It is decompofable at certain temperatures, by moft of the bodies poffeffing affinity for oxygene. It is foluble in about thirty times its bulk of water, and as far as we are acquainted with its affinities, incapable of

rate ; the oxygene forming the inferior, the nitrogene the fuperior ftratum.

' The fuppofition of the *chemical* compofition of atmof-pheric air, has been advanced by many philofophers. The two firft evidences have been often noticed.

combining with moſt of the ſimple and compound ſubſtances. 100 cubic inches of it
weigh about 31 grains at 55° temperature, and
30 atmoſpheric preſſure.

NITROUS OXIDE is a gas unalterable in its
conſtitution, at temperatures below ignition.
It is compoſed of oxygene and nitrogene, exiſting *perhaps* in the moſt intimate union which
thoſe ſubſtances are capable of aſſuming.*
Its properties approach to thoſe of acids. It
is decompoſable by the combuſtible bodies at
very high temperatures, is ſoluble in double its
volume of water, and in half its bulk of moſt of
the inflammable fluids. It is combinable with
the alkalies, and capable of forming with them
peculiar ſalts. 100 grains of it are compoſed
of about 63 nitrogene, and 37 oxygene.
100 cubic inches of it weigh 50 grains,
at 55° temperature, and 30 atmoſpheric preſſure.

---

* For it is unalterable by thoſe bodies which are capable
of attracting oxygene from nitrous gas and nitrous acid, at
common temperatures.

NITROUS GAS is compofed of about ,56 oxygene, and ,44 nitrogene, in intimate union. It is foluble in twelve times its bulk of water, and is combinable with the acids, and certain metallic folutions; it is poffeffed of no acid properties, and is decompofable by moft of the bodies that attract oxygene ftrongly, at high temperatures. 100 cubic inches of it weigh about 34 grains, at the mean temperature and preffure.

NITRIC ACID is a fubftance permanently aëriform at common temperatures, compofed of about 1 nitrogene, to 2,3 oxygene. It is foluble to a great extent in water, and combinable with the alkalies, and nitrous gas. It is decompofable by moft of the combuftible bodies, at certain temperatures. 100 cubic inches of it weigh, at the mean temperature and preffure, nearly 76 grains.

RESEARCH III.

RELATING TO THE RESPIRATION OF

NITROUS OXIDE,

AND OTHER

GASES.

RESEARCH III.

DIVISION I.

EXPERIMENTS and OBSERVATIONS on the EFFECTS produced upon ANIMALS by the RESPIRATION of NITROUS OXIDE.

I. *Preliminaries.*

THE term *respirable*, in its phyfiological application, has been differently employed. Some times by the refpirability of a gas has been meant, its power of fupporting life for a great length of time, when repeatedly applied to the blood in the lungs. At other times all gafes have been confidered as refpirable, which were capable of introduction into the lungs by voluntary efforts, without any relation to their vitality.

In the laſt ſenſe the word reſpirable is moſt properly employed. In this ſenſe it is uſed in the following ſections.

Non-reſpirable gaſes are thoſe, which when applied to the external organs of reſpiration, ſtimulate the muſcles of the epiglottis in ſuch a way as to keep it perfectly cloſe on the glottis; thus preventing the ſmalleſt particle of gas from entering into the bronchia, in ſpite of voluntary exertions; ſuch are carbonic acid, and acid gaſes in general.*

Of reſpirable gaſes, or thoſe which are capable of being taken into the lungs by voluntary efforts.

One only has the power of uniformly ſupporting life;—atmoſpheric air. Other gaſes, when reſpired, ſooner or later produce death; but in different modes.

Some, as nitrogene and hydrogene, effect no poſitive change in the venous blood. Animals

* See the curious experiments of Roſier, Journal de Phyſique, 1786, vol. 1, pag. 419.

immerſed in theſe gaſes die of a diſeaſe, produced by privation of atmoſpheric air, analogous
to that occaſioned by their ſubmerſion in water,
or non-reſpirable gaſes.

Others, as the different varieties of hydrocarbonate, deſtroy life by producing ſome poſitive change* in the blood, which probably immediately renders it incapable of ſupplying
the nervous and muſcular fibres with principles
eſſential to ſenſibility and irritability.

Oxygene, which is capable of being reſpired
for a much greater length of time than any
other gas, except common air, finally deſtroys
life ; firſt producing changes in the blood,
connected with new living action.†

After experiments, to be detailed hereafter,
made upon myſelf and others, had proved that
nitrous oxide was reſpirable, and capable of

___

* As appears from the experiments of Dr. Beddoes ;
likewiſe thoſe of Mr. Watt.

† As appears from the experiments of Lavoiſier and Dr.
Beddoes ; and as will be ſeen hereafter.

supporting life for a longer time than any of the gafes, except atmofpheric air and oxygene, I was anxious to afcertain the effects of it upon animals, in cafes where its action could be carried to a full extent ; and to compare the changes occafioned by it in their organs, with thofe produced by other powers.

II. *On the refpiration of Nitrous Oxide by warm-blooded Animals.*

The nitrous oxide employed in the following experiments, was procured from nitrate of ammoniac, and received in large jars, filled with water previoufly faturated with the gas. The animal was introduced into the jar, by being carried under the water ; after its introduction, the jar was made to reft on a fhelf, about half an inch below the furface of the water ; and the animal carefully fupported, fo as to prevent his mouth from refting in the water.

This mode of experimenting, either under water or mercury, is abfolutely neceffary, to afcertain with accuracy the effects of pure gafes on living beings. In fome experiments that I made on the refpiration of nitrous oxide, by animals that were plunged into jars of it opened in the atmofphere, and immediately clofed after their introduction, the unknown quantities of common air carried in, were always fufficient to render the refults perfectly inaccurate.

Animals fuffer little or nothing by being paffed through water.

That the phænomena in thefe experiments might be more accurately obferved, two or three perfons were always prefent at the time of their execution, and an account of them was noted down immediately after.

a. A ftout and healthy young cat, of four or five months old, was introduced into a large jar of nitrous oxide. For ten or twelve moments he remained perfectly quiet, and then began to make violent motions, throwing himfelf round the jar in every direction. In two minutes he appeared quite exhaufted, and funk quietly to the bottom of the jar. On applying my hand

X

to the thorax, I found that the heart beat with extreme violence ; on feeling about the neck, I could diftinctly perceive a ftrong and quick pulfation of the carotids. In about three minutes the animal revived, and panted very much ; but ftill continued to lie on his fide. His infpirations then became longer and deeper, and he fometimes uttered very feeble cries. In four minutes the pulfations of the heart appeared quicker and feebler. His infpirations were at long intervals, and very irregular ; in five minutes the pulfe was hardly perceptible ; he made no motions, and appeared wholly fenfelefs. After five minutes and quarter he was taken out, and expofed to the atmofphere before a warm fire. In a few feconds he began to move, and to take deep infpirations. In five minutes he attempted to rife on his legs ; but foon fell again, the extremities being flightly convulfed. In eight or nine minutes he was able to walk; but his motions were ftaggering and unequal, the right leg being convulfed, whilft the other was apparently ftiff and immoveable ; in about half an hour he was almost completely recovered.

b. A healthy kitten, of about six weeks old, was introduced into nitrous oxide. She very soon began to make violent exertions, and in less than a minute fell to the bottom of the receiver, as if apoplectic. At this moment, applying my hand to her side, I felt the heart beating with great violence. She continued gasping, with long inspirations, for three minutes and half; at the end of five minutes and half she was taken out completely dead.

c. Another kitten of the same breed was introduced into nitrous oxide, the day after. She exhibited the same phænomena, and died in it in about five minutes and half.

d. A small dog that had accidentally met with a dislocation of the vertebræ of the loins, and was in great pain, as manifested by his moaning and whining, was introduced into a large jar of nitrous oxide. He immediately became quiet, and lay on his side in the jar, breathing very deeply. In four minutes his respiration became noisy, and his eyes sparkled very much. I was not able to apply my hand to the thorax. In five minutes he appeared fenfelefs, and in feven minutes was perfectly dead.

e. A ftrong rabbit, of ten or twelve months old, was introduced into nitrous oxide. He immediately began to ftruggle very much, and in a minute fell down fenfelefs : in two minutes the legs became convulfed, and his infpirations were deep and noify : in lefs than five minutes he appeared perfectly dead.

f. A rabbit of a month old introduced into nitrous oxide, became fenfelefs in lefs than a minute ; the pulfations of the heart were very ftrong at this moment : they gradually became weaker, and in three minutes and half the animal was dead.

g. Another rabbit of the fame breed, after being rendered fenfelefs in nitrous oxide in a minute and half, was taken out. He foon became convulfed; in a minute began to breathe quickly ; in two minutes attempted to rife, but ftaggered, and fell again on his fide. His hinder legs were paralytic for near five minutes. In twenty he had almoſt recovered.

g. A middle ſized guinea-pig was much con-vulſed, after being in nitrous oxide for a minute. In two minutes and half he was fenſelefs. Taken out at this period, he remained for ſome minutes by the ſide of a warm fire, without moving; his fore legs then became convulſed; his hind legs were perfectly paralytic. In this ſtate he continued, without attempting to riſe or move, for near an hour, when he died.

h. A large and old guinea-pig died in nitrous oxide, exhibiting the ſame phænomena as the other animals, in about five minutes and quar-ter. A young one was killed in three minutes and half.

i. A ſmall guinea-pig, after breathing nitrous oxide for a minute and half, was taken out, and placed before a warm fire. He was for a few minutes a little convulſed; but in a quarter of an hour got quite well, and did not relapſe.

k. A large mouſe introduced into nitrous oxide, was for a few ſeconds very active. In half a minute he fell down fenfelefs ; in a minute and quarter he appeared perfectly dead.

l: A moufe taken out of nitrous oxide, after being in it for half a minute, continued convulfed for fome minutes, but finally recovered!

m. A young hen was introduced into a veffel filled with nitrous oxide. She immediately began to ftruggle very much ; fell on her breaft in lefs than half a minute, and in two minutes was quite dead.

n. A goldfinch died in nitrous oxide in lefs than a minute.

In each of thefe experiments a certain abforption of the gas was always perceived, the water rifing in the jar during the refpiration of the animal. From them we learn

1ft. That nitrous oxide is deftructive when refpired for a certain time to the warm blooded animals, apparently previoufly exciting them to a great extent.

2dly. That when its operation is ftopped before compleat exhauftion is brought on, the healthy living action is capable of being gradually reproduced, by enabling the animal to refpire atmof heric air.

3dly! That exhauftion, and death is produced in the fmall animals by nitrous oxide fooner than in the larger ones, and in young animals of the fame fpecies, in a fhorter time, than in old ones, as indeed Dr. Beddoes had conjectured a priori would be the cafe.

Moft of the animals deftroyed in thefe experiments were examined after death; the appearances in their organs were peculiar. To prevent unneceffary repetitions, an account of them will be given in the fourth fection.

III. *Effects of the respiration of Nitrous Oxide upon animals, as compared with thofe produced by their immerfion in Hydrogene and Water.*

Before the following experiments were made, a number of circumftances had convinced me that nitrous oxide acted on animals by producing fome pofitive change in their blood, connected with new living action of the irritable and fenfitive organs, and terminating in their death.

To afcertain however, the difference between the effects of this gas and thofe of hydrogene and non-refpirable gafes, I proceeded in the following way.

a. Of two healthy rabbits of about two months old, of the fame breed, and nearly of the fame fize.

One was introduced into nitrous oxide. In a half a minute, it had fallen down apparently fenfelefs. On applying my hand to the thorax, the action of the heart appeared, at firft, very quick and ftrong, it gradually became weaker, and in two minutes and half, the animal was taken out quite dead.

The other was introduced into a jar of pure hydrogene through water. He immediately began to ftruggle very much, and in a quarter of a minute fell on his fide. On feeling the thorax, the pulfations of the heart appeared very quick and feeble, they gradually diminifhed; his breathing became momentarily fhorter, and in rather more than three quarters of a minute, he was taken out dead. Dr. Kinglake was

present at this experiment, and afterwards
dissected both of the animals.

b. Of two similar rabbits of the same breed,
nearly three months old. One was introduced
into nitrous oxide, and after being rendered
senseless by the respiration of it for nearly a
minute and half, was exposed to the atmosphere, before a warm fire. He recovered gradually, but was occasionally convulsed, and had
a paralysis of one of his hinder legs for some
minutes : in an hour he was able to walk. The
other, after being immerged in hydrogene, for
near half a minute, was restored to the atmosphere apparently inanimate. In less than a
minute he began to breathe, and to utter a
feeble noise ; in two minutes was able to walk,
and in less than three minutes appeared perfectly
recovered.

b. A kitten of about two months old, was
introduced into a jar of nitrous oxide, at the
same time that another of the same breed was
plunged under a jar of water. They both
struggled very much. The animal in the nitrous

oxide fell fenfelefs before that under water had ceafed to ftruggle, and to throw out air from its lungs. In two minutes and three quarters, the animal under water was quite dead : it was taken out and expofed to heat and air, but did not fhew the flighteft figns of life. At the end of three minutes and half, the animal in nitrous oxide began to gafp, breathing very flowly; at four minutes and three quarters it was yet alive; at the end of five minutes and quarter it appeared perfectly dead. It was taken out, and did not recover.

From thefe experiments it was evident, that animals lived at leaft twice as long in nitrous oxide as in hydrogene or water. Confequently from this circumftance alone, there was every reafon to fuppofe that their death in nitrous oxide could not depend on the fimple privation of atmofpheric air; but that it was owing to fome peculiar changes effected in the blood by the gas.

IV. *Of the changes effected in the Organifation of warm-blooded Animals, by the refpiration
of Nitrous Oxide.*

The external appearance of animals that have
been deftroyed in nitrous oxide, is very little
different from that of thofe killed by privation
of atmofpheric air. The fauces and tongue
appear of a dark red, and the eyes are dull, and
a little protruded. Their internal organs, however, exhibit a very peculiar change. The
lungs are pale brown red, and covered here
and there with purple fpots ; the liver is of a
very bright red, and the mufcular fibre in
general dark. Both the auricles and ventricles
of the heart are filled with blood. The auricles
contract for minutes after the death of the animal.
The blood in the left ventricle, and the aorta, is of
a tinge between purple and-red, whilft that in
the right ventricle is of a dark color, rather more
purple than the venous blood. But thefe appearances, and their caufes, will be better underftood
after the following comparative obfervations are
read.

a. Of two fimilar rabbits, about eight months old, one A, was killed by expofure for near fix minutes to nitrous oxide, the other, B, was deftroyed by a blow on the head.

, They were both opened as fpeedily as poffible. The lungs of B were pale, and uniform in their appearance; this organ in A was redder, and every where marked with purple fpots. The liver of A was of a dark and bright red, that of B of a pale red brown. The diaphragm of B, when cut, was ftrongly irritable; that of A rather darker, and fcarce, at all contractile. All the cavities of the heart contracted for more than 50 minutes in B. The auricles contracted for near 25 minutes with force and velocity in A: but the ventricles were almoft inactive. The vena cava, and the right auricle, in A, were filled with blood, apparently a fhade darker than in B. The blood in the left auricle, and the aorta, appeared in A of a purple, a fhade brighter than that of the venous blood. In the left auricle of B it was red.

I opened the head of each, but not without injuring the brains, fo that I was unable to make any accurate comparifon. The color of the brain in A appeared rather darker than in B.

b. Two rabbits, C and D, were deftroyed, C by immerfion in nitrous oxide, D in hydrogene: they were both diffected by Dr. Kinglake. The blood in the pulmonary vein and the left auricle of C was of a different tinge, from that in D more inclined to purple red. The membrane of the lungs in C was covered with purple fpots, that of D was pale and uniform in its appearance. The brain in C was rather darker than in D; but there was no perceptible effufion of blood into the ventricles either in D or C. The liver in C was of a brighter red than in health, that in D rather paler.

c. In the laft experiment, the comparative irritability of the ventricles and auricles of the heart and the mufcular fibre in the two animals, had not been examined. That thefe circumftances might be noticed, two rabbits, E and F were killed; E under water in about a minute, and F in nitrous oxide in three minutes. They were immediately opened, and after a minute, the appearance of the heart, and organs of refpiration obferved.

Both the right and left ventricles of the heart in F contracted but very feebly; the auricles regularly and quickly contracted; the aorta appeared perfectly full of blood. In E, a feeble contraction of the left finus venosus and auricle was obferved; the left ventricle did not contract: the right contracted, but more flowly than in F. In a few minutes, the contractions of the ventricles in F had ceafed, whilst the auricles contracted as ftrongly and quickly as before. The blood in the pulmonary veins of F was rather of a redder purple than in E; the difference of the blood in the vena cava was hardly perceptible, perhaps it was a little more purple in F. The membranous fubftance of the lungs in F was fpotted with purple as from extravafated blood, whilft that in E was pale. The brain in F was darker than in E. On opening the ventricles no extravafation of blood was perceptible.

The auricles of the heart in F contracted ftrongly for near twenty minutes; and then gradually their motion became lefs frequent; in twenty-eight minutes it had wholly ceafed. The right auricle and ventricle in E, occafionally contracted for half an hour. The livers of both animals were fimilar when they were firft opened, of a dark red; that of F preferved its color for fome time when expofed to the atmofphere; whilft that of E almoft immediately became paler under the fame circumftances.

The periftaltic motion continued for nearly an equal time in both animals.

d. The fternum of a young rabbit was removed fo that the heart and lungs could be perceived; and he was introduced into a veffel filled with nitrous oxide; the blood in the pulmonary veins gradually became more purple, and the heart appeared to beat quicker than before, all the mufcles contracting with great force. After he had been in about a minute, fpots began to appear on the lungs, though the contractions of. the heart; became quicker and weaker; in three minutes and half he was quite dead; after death the ventricles contracted very feebly, though the contractions of the auricles were as strong almost after the end of five minutes as at firſt. This animal was paſſed through water saturated with nitrous oxide; poſſibly this fluid had some effect on his organs.

Beſides theſe animals, many others, as guinea-pigs, mice and birds, were diſſected, after being deſtroyed in nitrous oxide; in all of them the ſame general appearance was obſerved. Their muſcular fibre almoſt always appeared leſs irritable than that of animals deſtroyed, by organic læſion of part of the nervous ſyſtem, in the atmoſphere. The ventricles of the heart in general, contracted feebly and for a very ſhort time; whilſt the auricles continued to act for a great length of time. The lungs were dark in their appearance, and always ſuffuſed here and there with purple; the blood in the pulmonary veins when ſlightly obferved, appeared dark, like venous blood, but when minutely examined, was evidently much more purple. The blood in the vena cava, was darker than that in the pulmonary veins. The cerebrum was dark.

In a late experiment, I thought I perceived a flight extravafation of blood in one of the ventricles of the brain in a rabbit deftroyed in nitrous oxide ; but as this appearance had not occurred in the animals I had examined before, or in thofe diffeded by Dr. Kinglake, and Mr. King, Surgeon, I am inclined to refer it to an accidental caufe. At my requeft, Mr. Smith, Surgeon, examined the brain of a young rabbit that had been killed in his prefence in nitrous oxide ; he was of opinion that no effufion of blood into the ventricles had taken place.

In comparing the external appearance of the crural nerves in two rabbits that had been diffeded by Dr. Kinglake, having been deftroyed one in hydrogene, the other in nitrous oxide, we could perceive no perceptible difference.

It deferves to be noticed, that whenever the

gall bladder and the urinary bladder have been examined in animals deftroyed in nitrous oxide, they have been always diftended with fluid; which is hardly ever the cafe in animals killed by privation of atmofpheric air.

In the infancy of my experiments on the action of nitrous oxide upon animals, I thought that it rendered the venous blood lefs coagulable; but this I now find to be a miftake. The blood from the pulmonary veins of animals killed in nitrous oxide, does not fenfibly differ in this refpect from the arterial blood of thofe deftroyed in hydrogene, and both become vermilion nearly in the fame time when expofed to the atmofphere.

In defcribing the various fhades of color of the blood in the preceding obfervations on the different diffected animals, the poverty of the language of color, has obliged me to adopt terms, which I fear will hardly convey to the mind of the reader, diftinct notions of the differences obfervable by minute examination in the venous and arterial blood of

animals that die of privation of atmofpheric
air, and of thofe deftroyed by the action of
nitrous oxide. This difference can only be
obferved in the veffels by means of a ftrong
light ; it may however be eafily noticed in the
fluid blood by the introduction of it from the
arteries or veins at the moment of their inci-
fion, between two polifhed furfaces of white
glafs,* fo clofely adapted to each other, as to
prevent the blood from coming in contact with
the atmofphere.

Having four or five times had an opportunity
of bleeding people in the arm for trifling complaints, I have always received the blood in
phials, filled with various gafes, in a mode
to be defcribed hereafter. Venous blood
agitated in nitrous oxide, compared with fimilar blood in common air, hydrogene, and nitrogene, was always darker and more purple

*The colour of common venous blood, examined in
this way, refembles that of the paint called by colour-men
red ochre ; that of blood faturated with nitrous oxide, ap-
proaches to the tinge of lake.

than the firſt, and much brighter and more
florid than the two laſt, which were not different in their color from venous' blood, received
between two ſurfaces of glaſs. It will be ſeen
hereafter, that the coagulum of venous blood
is rendered more purple when expoſed to nitrous oxide, whilſt the gas is abſorbed; likewiſe
that blood altered by nitrous oxide, is capable
of being again rendered vermilion, by expoſure to the air.

The appearances noticed in the above mentioned experiments, in the lungs of animals
deſtroyed in nitrous oxide, are ſimilar to thoſe
obſerved by Dr. Beddoes, in animals that had
been made to breathe oxygene for a great
length of time.

There were many reaſons for ſuppoſing that
the large purple ſpots in the lungs of animals
deſtroyed in nitrous oxide, were owing to ex-
travaſation of venous blood from the capillary
veſſels; their coats being broken by the highly
increaſed arterial action. To aſcertain whether
theſe phænomena exiſted at a period of the

action of nitrous oxide, when the animal was recoverable by expofure to the atmofphere,

I introduced a rabbit of fix months old, into a veffel of nitrous oxide, and after a minute, when it had fallen down apparently apoplectic, plunged him wholly under water; he immediately began to ftruggle, and what furprifed me very much, died in lefs than a minute after fubmerfion. On opening the thorax, the blood in the pulmonary veins was nearly of the color of that in animals that have been fimply drowned: The lungs were here and there, marked with a few points; but there were no large purple fpots, as in animals that have been wholly deftroyed in nitrous oxide: the right fide of the heart only contracted. In this experiment, the excitement from the action of the gas was probably carried to fuch an extent, as to produce indirect debility. There are reafons for fuppofing, that animals after having been excited to but a fmall extent by the refpiration of nitrous oxide, will live under water for a greater length of time, than animals previoufly made to breathe common air.

V. *Of the respiration of mixtures of Nitrous Oxide, and other gases, by warm-blooded Animals.*

a. A rabbit of near two months old, was introduced into a mixture of equal parts hydrogene and nitrous oxide through water. He immediately began to struggle ; in a minute fell on his side ; in three minutes gasped, and made long inspirations ; and in four minutes and half, was dead. On diffection, he exhibited the fame appearances as animals deftroyed in nitrous oxide.

b. A large and ftrong moufe was introduced into a mixture of three parts hydrogene to one part nitrous oxide. He immediately began to ftruggle very much, in half a minute; became convulfed, and in about a minute, was quite dead.

c. Into a mixture of one oxygene, and three nitrous oxide, a fmall guinea-pig was introduced. He immediately began to ftruggle, and in two minutes repofed on his fide, breathing very deeply. He made afterwards no violent mufcular motion ; but lived quietly for near fourteen minutes : at the end of which time, his legs were much convulfed. He was taken out, and recovered.

d. A moufe lived apparently without fuffer- ing, for near ten minutes, in a mixture of 1 atmofpheric air, and 3 nitrous oxide, at the end of eleven minutes he began to ftruggle, and in thirteen minutes became much convulfed.

e. A cat of three months old, lived for feven- teen minutes, in a very large quantity of a mix- ture of 1 atmofpheric air, and 12 nitrous oxide. On her firft introduction fhe was very much agitated and convulfed, in a minute and half fhe fell down as if apoplectic, and continued breathing very deeply during the remainder of the time, fometimes uttering very feeble cries. When taken out, fhe appeared almoft inanimate, but on being laid before the fire, gradually began to breathe and move ; being for fome time, like moft of the animals that have recovered after breathing nitrous oxide, con- vulfed on one fide, and paralytic the other.

f. A goldfinch lived for near five minutes in a mixture of equal parts nitrous oxide and oxygene, without apparently suffering. Taken out, he appeared faint and languid, but finally recovered.*

VI. *Recapitulation of facts relating to the respiration of Nitrous Oxide, by warm-blooded Animals.*

1. Warm-blooded animals die in nitrous oxide infinitely sooner than in common air or oxygene; but not nearly in so short a time as in gases incapable of effecting positive changes in the venous blood, or in non-respirable gases.

2. The larger animals live longer in nitrous oxide than the smaller ones, and young animals

* Small birds suffer much from cold when introduced into gases through water. In this experiment, the goldfinch was immediately inserted into a large mouthed phial, filled with the gases, and opened in the atmosphere.

die in it fooner than old ones of the fame fpecies.

.3. When animals, after breathing nitrous oxide, are removed from it before compleat exhauftion has taken place, they are capable of being reftored to health under the action of atmofpheric air.

4. Peculiar changes are effected in the organs of animals by the refpiration of nitrous oxide. In animals deftroyed by it, the arterial blood is purple red, the lungs are covered with purple fpots, both the hollow and compact mufcles are *apparently* very inirritable, and the brain is dark colored.

5. Animals are deftroyed by the refpiration of mixtures of nitrous oxide and hydrogene nearly in the fame time as by pure nitrous ox-ide; they are capable of living for a great length of time in nitrous oxide mingled with very mi-nute quantities of oxygene or common air.

Thefe facts will be reafoned upon in the next divifion.

VII. *Of the respiration of Nitrous Oxide by amphibious Animals.*

As from the foregoing experiments, it appeared that the nitrous oxide deftroyed warm-blooded animals by increafing the living action of their organs to fuch an extent, as finally to exhauft their irritability and fenfibility ; it was reafonable to conjecture that the cold-blooded animals, poffeffed of voluntary power over refpiration, would fo regulate the quantity of nitrous oxide applied to the blood in their lungs as to bear its action for a great length of time. This conjecture was put to the teft of experiment ; the following facts will prove its error.

a. Of two middle-fized water-lizards, one was introduced into a fmall jar filled with nitrous oxide, over moift mercury, by being paffed through the mercury ; the other was made to breathe hydrogene, by being carried into it in the fame manner.

The lizard in nitrous oxide, in two or three minutes, began to make violent motions, appeared very uneafy, and rolled about the jar in every direction, fometimes attempting to climb to the top of it. The animal in hydrogene was all this time very quiet, and crawled about the veffel without being apparently much affected. At the end of twelve minutes, the lizard in nitrous oxide was lying on his back feemingly dead; but on agitating the jar he moved a little; at the end of fifteen minutes he did not move on agitation, and his paws were refting on his belly. He was now taken out ftiff and apparently life-lefs, but after being expofed to the atmofphere for three or four minutes, took an infpira-tion, and moved his head a little; he then raifed the end of his tail, though the middle of it was ftill ftiff and did not bend when touched. His four legs remained clofe to his fide, and were apparently ufelefs; but on pricking them with the point of a lancet, they became con-vulfed. After being introduced into fhallow water, he was able to crawl in a quarter of an hour, though his motions were very irre-gular. In an hour he was quite well. The animal in hydrogene appeared to have suffered very little in three quarters of an hour, and had raised himself against the side of the jar. At the end of an hour he was taken out, and very soon recovered.

b. Some hours after, the same lizards were again experimented upon. That which had been inserted into hydrogene in the last experiment, being now inserted into nitrous oxide.

This lizard was apparently lifeless in fourteen minutes, having tumbled and writhed himself very much during the first ten minutes. Taken out after being in twenty-five minutes, he did not recover. The other lizard lived in hydrogene for near an hour and quarter, taken out after an hour and twenty minutes, he was dead.

These animals were both opened, but the viscera of the nitrous oxide lizard were so much injured by the knife, that no accurate comparison of them with those of the other could be made, I thought that the lungs appeared rather redder.

c. Of two fimilar large water-lizards, one was introduced into a veffel ftanding over mercury, wholly filled with water that had been long boiled and fuffered to cool under mercury.

The animal very often rofe to the top of the jar as, if in fearch of air, during the firft half hour; but fhewed no other figns of uneafinefs. At the end of three quarters of an hour, he became very weak, and appeared fcarcely able to fwim in the water. Taken out at the end of fifty minutes, he recovered.

The other was inferted into nitrous oxide. After much ftruggling, he became fenfelefs in about fifteen minutes, and lay on his back. Taken out at the end of twenty minutes, he remained for a long time motionlefs and ftiff, but in a quarter of an hour, began to move fome of his limbs.

From thefe experiments, we may conclude, that water-lizards, and moft probably the other amphibious animals, die in nitrous oxide in a much fhorter time than in hydrogene or pure water; confequently their death in it cannot depend on the fimple privation of atmofpheric air.

At the feafon of the year in which this inveftigation was carried on, I was unable to
procure frogs or toads. This I regret very much.

Suppofing that cold-blooded animals die
in nitrous oxide from pofitive changes effected in their blood by the gas, it would be
extremely interefting to notice the apparent alterations taking place in their organs of refpiration and circulation during its action, which
could eafily be done, the membranous fubftance
of their lungs being tranfparent. The increafe or
diminution of the irritability of their mufcular
fibre, might be determined by comparative galvanic experiments.

VIII. *Effects of folution of nitrous oxide in
water on Fifhes.*

. *a.* A fmall flounder was introduced into a
veffel filled with folution of nitrous oxide in
water over mercury. He remained at reft for
ten minutes and then began to move about the

jar in different directions. In a half an hour he was apparently dying, lying on his side in the water. He was now taken out, and introduced into a veffel filled with water faturated with common air, he very foon recovered.

b. Of two large thornbacks,* equally brifk and lively. One, A, was introduced into a jar containing nearly 3 cubic inches of water, faturated with nitrous oxide, and which pre-vious to its impregnation had been long boiled ; the other, B, was introduced into an equal quantity of water which had been deprived of air by diftillation through mercury.

A, appeared very quiet for two or three minutes, and then began to move up and down in the jar, as if agitated. In eight minutes his motions became very irregular, and he darted obliquely from one fide of the jar to the other;

---

* I ufe the popular name. This fifh is very common in every part of England; it is nearly of the fame fize and color as the minnow, and is diftinguifhed from it by two fmall bony excrefences at the origin of the belly. It is extremely fufceptible.

In twelve minutes, he became ſtill, and moved his gills very ſlowly. In fifteen minutes he appeared dead. After ſixteen minutes he was taken out; but ſhewed no ſigns of life.

B was very quiet for four minutes and half. He then began to move about the jar. In ſeven minutes he had fallen on his back, but ſtill continued to move his gills. In eleven minutes he was motionleſs; taken out after thirteen minutes, he did not recover.

c. Of two thornbacks, one, C was introduced into about an ounce of boiled water in contact with hydrogene, ſtanding over mercury. The other, D, was introduced into well boiled water ſaturated with nitrous oxide, and ſtanding in contact with it over mercury. C lived near thirteen minutes, and died without being previouſly much agitated. D was apparently motionleſs, after having the ſame affections as A in the laſt experiment, in ſixteen minutes. At the end of this time he was taken out and introduced into common water. He ſoon began to move his gills, and in leſs than a quarter of an hour was fo far recovered as to be able to fwim.

The laft experiment was repeated on two fmaller thornbacks; that in the aqueous folution of nitrous oxide lived near feventeen minutes, that in the water in contact with hydrogene, about fifteen and half.

The experiments in Ref. I. Div. 3, prove the difficulty, and indeed almoft impoffibility of driving from water by boiling, the whole of the atmofpheric air held in folution by it; they likewife fhow that nitrous.oxide. by its ftrong affinity for water, is capable of expelling air from that fluid after no more can be procured from it by ebullition.

Hence, if water faturated with nitrous oxide had no pofitive effects upon fifhes; they ought to die in it much fooner than in water deprived of air by ebullition. From their living in it rather longer;* we may conclude, that they are deftroyed not by privation of atmofpheric air, but

---

* A priori I expected that fifhes, like amphibious animals would have been very quickly deftroyed by the action of nitrous oxide.

from some positive change effected in their blood by the gas.

A long while ago, from observing that the gills of fish became rather of a lighter red during their death, in the atmosphere; I conjectured that the disease of which they died, was probably hyperoxygenation of the blood connected with highly increased animal heat. For not only is oxygene presented to their blood in much larger quantities in atmospheric air than in its aqueous solution; but likewise, to use common language, in a state in which it contains much more *latent heat*. Without however laying any stress on this supposition, I had the curiosity to try whether thornbacks would live longest in atmospheric air or nitrous oxide. In one experiment, they appeared to die in them nearly in the same time. In another, the fish in nitrous oxide lived nearly half as long again as that in atmospheric air.

XI. *Effects of Nitrous Oxide on Insects.*

The winged insects furnished with breathing

holes, become motionlefs in nitrous oxide very fpeedily ; being however poffeffed of a certain voluntary power over refpiration, they fometimes recover, after having been expofed to it for fome minutes, under the action of atmofpheric air.

A butterfly was introduced into a fmall jar, filled with pure nitrous oxide, over mercury. He ftruggled a little during the firft two or three feconds ; in about feven feconds, his legs became convulfed, and his wings were wrapt round his body ; in about half a minute he was fenfelefs ; taken out after fix minutes, he did not recover.

Another butterfly introduced into hydrogene, became convulfed in about a quarter of a minute, was fenfelefs in twenty feconds, and taken out after five minutes, did not revive.

A large drone, after being in nitrous oxide for a minute and a quarter, was taken out fenfelefs. After being for fome time expofed to the atmofphere, he began to move, and at laft rofe on his wings. For fome time, however, he was unable to fly in a ftraight line ; and often after describing circles in the air, fell to the ground as if giddy.

A large fly, became motionlefs in nitrous oxide after being convulfed, in about half a minute. Another was rendered fenfelefs in hydrogene, in lefs than a quarter of a minute.

A fly introduced into hydrocarbonate, dropt immediately fenfelefs ; taken out after about a quarter of a minute, he recovered ; but like the fly that had lived after expofure to nitrous oxide, was for fome time vertiginous.

Flies live much longer under water, alcohol, or oil, than in non-refpirable gafes, or gafes incapable of fupporting life. A certain quantity of air always continues attached in the fluid to the fine hairs furrounding their breathing holes, fufficient to fupport life for a fhort time.

Snails and earth-worms, live in nitrous oxide a long while, they die in it however, much fooner than in water or hydrogene ; probably from the fame caufes as the amphibious animals.

DIVISION II.

Of the CHANGES *effected in* NITROUS OXIDE, *and other* GASES, *by the* RESPIRATION *of* ANIMALS.

─────────

I. *Preliminaries.*

As foon as I had difcovered that nitrous oxide was refpirable, and poffeffed of extraordinary powers of action on living beings, I was anxious to be acquainted with the changes effected in it by the venous blood. To inveftigate thefe changes, appeared at firft a fimple problem; I foon however found that it involved much preliminary knowledge of the chemical proper-ties and affinities of nitrous oxide. After I had afcertained by experiments detailed in the pre-ceding Refearches, the compofition of this gas its combinations, and the phyſical changes effected by it in living beings, I began my enquiry relating to the mode of its operation.

Finding that the reſidual gas of nitrous oxide after it had been breathed for ſome time in ſilk bags, was chiefly nitrogene, I at firſt conjectured that nitrous oxide was decompoſed in reſpiration in the ſame manner as atmoſpheric air, and its oxygene only combined with the venous blood; the following experiments ſoon however convinced me of my error.

II. *Abſorption of Nitrous Oxide by venous blood. Changes effected in the blood by different Gaſes.*

a. Though the laws of the coagulability of the blood are unknown, yet we are certain that at the moment of coagulation, a perfectly new arrangement of its principles takes place; conſequently, their powers of combination muſt be newly modified. The affinities of living blood can only be afcertained during its circulation in the veffels of animals. At the moment of effufion from thofe veffels, it begins to pafs through a feries of changes, which firft produce coagulation, and finally its compleat decompofition.

Confequently, the action of fluid blood upon gafes out of the veffels, will be more analogous to that of circulating blood in proportion as it is more fpeedily placed in contact with them.

b. To afcertain the changes effected in nitrous oxide by fluid venous blood.

A jar, fix inches long and half an inch wide, graduated to ,05 cubic inches, having a tight ftopper adapted to it, was filled with nitrogene; which is a gas incapable of combining with, and poffeffing no power of action upon venous blood. A large orifice was made in the vein of a tolerably healthy man, and the ftopper removed from the jar, which was brought in contact with the arm fo as to receive the blood, and preffed clofe againft the fkin, in fuch a way as to leave an orifice juft fufficient for the efcape of the nitrogene, as the blood flowed in. When the jar was full, it was clofed, and carried to the pneumatic apparatus, the mercury of which had been previoufly a little warmed. A fmall quantity of the blood was transferred into another jar to make room for the gas. The remaining quantity equalled exactly two cubic inches ; to this was introduced as fpeedily as poffible, eleven meafures equal to ,55 cubic inches of nitrous oxide, which left a refiduum of $\frac{1}{32}$ only, when abforbed by boiled water, and was confequently, perfectly pure. On agitation, a rapid diminution of the gas took place.

In the mafs of blood which was opaque, but little change of color could be perceived ; but that portion of it diffufed over the fides of the jar, was evidently of a brighter purple than the venous blood.

It was agitated for two or three minutes, and then fuffered to reft ; in eight minutes it had wholly coagulated ; a fmall quantity of ferum had feparated, and was diffufed over the coagulum. This coagulum was dark ; but evidently of a more purple tinge than that of venous blood ; no gas had apparently been liberated during its formation.

The nitrous oxide remaining, was not quite equal to feven meafures; hence, at leaft four meafures of it had been abforbed.

To afcertain the nature of the refiduum, it was neceffary to transfer it into another veffel, but this I found very difficult to accomplifh, on account of the coagulated blood. By piercing through the coagulum and removing part of it by means of curved iron forceps, I at laft contrived to introduce about $4\frac{1}{2}$ meafures of the gas into a fmall cylinder, graduated to ,25 cubic inches, in which it occupied of courfe, nearly 9 meafures ; when a little folution of ftrontian was admitted to thefe, it became very flightly clouded ; but the abforption that took place did not more than equal half its bulk. Confequently, the quantity of carbonic acid evolved from the blood, or formed, muft have been extremely minute.

On the introduction of pure water, a rapid

abforption of the gas took place, and after agitation, not quite 3 meafures remained. Thefe did not *perceptibly* diminifh with nitrous gas; their quantity was too fmall to be examined by any other teft ; but there is reafon to fuppofe that they were chiefly compofed of nitrogene.

From this experiment, it appeared that nitrous oxide is abforbed when placed in contact with venous blood ; at the fame time, that a very minute quantity of carbonic acid and probably nitrogene is produced.

c. In another fimilar experiment when nearly half a cubic inch of nitrous oxide was abforbed by about a cubic inch and three quarters of fluid blood, the refidual gas did not equal more than $\frac{1}{8}$, the quantity abforbed being taken as unity. This fact induced me to fuppofe that the abforption of nitrous oxide by venous blood, was owing to a fimple folution of the gas in that fluid, analogous to its folution in water or alcohol.

To afcertain if nitrous oxide could be expelled from blood impregnated with it, by heat ; I introduced to 2 cubic inches of fluid blood taken from the medial vein, about ,6 cubic inches of nitrous oxide. After agitation, in. feven minutes nearly ,4 were abforbed. In ten minutes, after the blood had completely coagulated, the cylinder containing it, was transferred in contact with mercury, into a veffel of folution of falt in water ; this folution was heated and made to boil. During its ebullition, the whole of the blood became either white or pale brown, and formed a folid coherent mafs ; whilft fmall globules of gas were given out from it. In a few minutes, about ,25 of gas had collected. After the veffel had cooled, I attempted to transfer this gas into a fmall graduated jar in the mercurial apparatus, but in vain ; the mafs in the jar was fo folid and tough, that I could not remove it. By tranf-ferring it to the water apparatus, I fucceeded in difplacing enough of the coagulum to fuffer the water to come in contact with the gas ; an abforption of nearly half of it took place ; hence, *I conjecture,* that nitrous oxide had been given out by the impregnated blood.

d. Some frefh dark coagulum of venous blood, was expofed to nitrous oxide. A very flight alteration of color took place at the furface of the blood, perceptible only in a ftrong light, and a minute quantity of gas was abforbed. A taper burnt in the remaining gas as brilliantly as before, hence, it had apparently fuffered no alteration.

e. To compare the phyfical changes effected in the venous blood by nitrous oxide, with thofe produced by other gafes, I made the following experiments.—I filled a large phial, containing near 14 cubic inches, with blood from the vein of the arm of a man, and immediately transferred it to the mercurial apparatus. Different portions of it were thrown into fmall graduated cylinders, filled with the following gafes : nitrogene, nitrous gas, common air, oxygene, nitrous oxide, carbonic acid, and hydrocarbonate.

The blood in each of them was fucceffively agitated till it began to coagulate ; and making allowances for the different periods of agitation, there was no marked difference in the times of coagulation.

The color of the coagulum in every part of the cylinder, containing nitrogene, was the fame very dark red. When it was agitated fo as to tinge the fides of the jar, it appeared exactly of the color of venous blood received between two furfaces of glafs ; no perceptible abforption of the gas had taken place.

The blood in nitrous gas was dark, and much more purple on the top than that in nitrogene. When agitated fo as to adhere to the jar as a thin furface, this purple was evidently deep and bright. An abforption of rather more than $\frac{1}{8}$ of the volume of gas had taken place.

The blood in oxygene and atmofpheric air, were of a much brighter tinge than that in any of the other gafes. On the top, the color was vermilion, but no perceptible abforption had taken place.

The coagulum in nitrous oxide, when examined in the mafs was dark, and hardly diftinguifhable in its color from venous blood ; but when minutely noticed at the furface where it was covered with ferum, and in its diffufion over the fides of the jar, it appeared of a fine purple red, a tinge brighter than the blood in nitrous gas. An abforption had taken place in this cylinder, more confiderable than in any of the others.

In carbonic acid, the coagulum was of a brown red, much darker than the venous blood, and a flight diminution of gas had taken place.

In the hydrocarbonate,* the blood was red, a fhade darker than the oxygenated blood, and a very flight diminution of the gas† was perceptible.

f. To human blood that had been faturated with nitrous oxide whilft warm and conftantly agitated for four or five minutes, to prevent its uniform coagulation, oxygene was introduced; the red purple on the furface of it,

---

* The hydrocarbonate employed, was procured from alcohol, by means of fulphuric acid. This gas contains more carbon, than hydrocarbonate from water and charcoal.

† The curious fact of the reddening of venous blood by hydrocarbonate, was difcovered by Dr. Beddoes.

immediately changed to vermilion; and on agitation, this color was diffused through it. On comparing the tinge with that of oxygenated blood, no perceptible difference could be observed. No change of volume of the oxygene introduced, had taken place; and consequently, no nitrous oxide had been evolved from the blood.

g. Blood, impregnated with nitrous gas, was exposed to oxygene; but after agitation in it for many minutes, no change of its dark purple tinge could be observed, though a flight diminution of the oxygene appeared to take place.

h. Blood that had been rendered vermilion in every part by long agitation in atmofpheric air, the coagulum of which was broken and diffufed with the coagulable lymph through the ferum, was exposed to nitrous oxide; for fome minutes no perceptible change of color took place; but by agitation for two or three hours, it evidently affumed a purple tinge, whilft a a flight abforption of gas took place. It never

however, became nearly fo dark as venous
blood that had been expofed to nitrous oxide.

i. Blood, oxygenated in the fame manner
as in the laft experiment, the coagulum of
which had been broken, was expofed to nitrous
gas. The furface of it immediately became
purple, and by agitation for a few minutes, this
color was diffufed through it. A flight diminution of the gas was obferved. On comparing
the tinge with that of venous blood that had
been previoufly expofed to nitrous gas, there
was no perceptible difference.

k. Blood expofed to oxygenated muriatic acid
is wholly altered in its conftitution and phyfical
properties, as has been often noticed ; the
coagulum becomes black in fome parts, and
brown and white in others. Venous blood, after
agitation in hydrogene or nitrogene, oxyge-
nates when expofed to the atmofphere in the
fame manner as fimple venous blood. I had
the curiofity to try whether venous blood
expofed to hydrogene, would retain its power
of being oxygenated longer than blood

faturated with nitrous oxide : for this purpofe fome fimilar black coagulum was agitated for fometime in two phials, one filled with hydrogene, the other with nitrous oxide. They were then fuffered to reft for three days at a temperature from about 56° to 63o. After being opened, no offenfive fmell was perceived in either of them, the blood in hydrogene was rather darker than at the time of their expofure, whilft that in nitrous oxide was of a brighter purple. On being agitated for fome time in the atmofphere, the blood in nitrous oxide became red, but not of fo bright a tinge as oxygenated venous blood. The color of the blood in hydrogene did not at all alter.

l. To afcertain whether impregnation with nitrous oxide accelerated or retarded the putrefaction of the blood ; I expofed venous blood in four phials, the firft filled with hydrocarbonate, the fecond with hydrogene ; the third with atmofpheric air, and the fourth with nitrous oxide. Examined after a fortnight, the blood in hydrogene and common air were both black, and ftunk very much ; that in hydrocarbonate was red, and perfectly fweet ; that in nitrous oxide appeared purple and had no difagreeable fmell.

In a fecond experiment, when blood was expofed for three weeks to hydrocarbonate and nitrous oxide, that in nitrous oxide was darker than before and ftunk a little ; that in hydrocarbonate was ftill perfectly fweet. The power of hydrocarbonate to prevent the putrefaction of animal matters, was long ago noticed by Mr. Watt.

m. Having accidentally cut one of my fingers fo as to lay bare a little mufcular fibre, I introduced it whilft bleeding into a bottle of nitrous oxide ; the blood that trickled from the wound evidently became much more purple ; but the pain was neither alleviated or increafed. When however, the finger was taken out of the nitrous oxide and expofed to the atmofphere, the wound fmarted more than it had done before. After it had ceafed to bleed, I inferted it through water into a veffel of nitrous gas ; but it did not become more painful than before.

From all these obfervations, we may conclude,

1ft. That when nitrous oxide is agitated in fluid venous blood, a certain portion of the gas is abforbed; whilft the color of the blood changes from dark red to red purple.

2dly. That during the abforption of nitrous oxide by the venous blood, minute portions of nitrogene and carbonic acid are produced, either by evolution from the blood, or from a decompofition of part of the nitrous oxide.

3dly. That venous blood impregnated with nitrous oxide is capable of oxygenation; and vice verfa; that oxygenated blood may be combined with nitrous oxide.

When blood feparated into coagulum and ferum, is expofed to nitrous oxide, it is moft probable that the gas is chiefly abforbed by the ferum. That nitrous oxide however is capable of acting upon the coagulum, is evident from d. In the fluid blood, as we fhall fee hereafter, nitrous oxide is abforbed by the attractions of the whole compound.

To afcertain whether the changes effected in nitrous oxide by the circulating blood acting through the moift coats of the pulmonary veins of living animals, were highly analogous to thofe produced in it by fluid venous blood removed from the veffels, I found extremely difficult.

I have before obferved, that when animals are made to refpire nitrous oxide, a certain abforption of the gas always takes place ; but the fmaller animals, the only ones that can be experimented upon in the mercurial apparatus, die in nitrous oxide fo fpeedily and occafion fo flight a diminution of gas, that I judged it ufelefs to attempt to analife the refiduum of their refpiration, which fupports flame as well as pure nitrous oxide, and is chiefly abforbable by water.

In the infancy of my refearches, I often refpired nitrous oxide in a large glafs bell, fur-

nifhed with a breathing tube and ftopcock, and
poifed in water faturated with the gas.

In two or three experiments, in which the noftrils being clofed after the exhauftion of the
lungs, the gas was infpired from the bell and
refpired into it, a confiderable diminution was
perceived, and by the teft of lime water fome
carbonic acid appeared to have been formed ;
but on account of the abforption of this carbonic acid by the impregnated water, and the
liberation of nitrous oxide from it, it was impoffible to determine with the leaft accuracy,
the quantities of products after refpiration.

About this time likewife, I often examined
the refiduum of nitrous oxide, after it had been
refpired in filk bags. In thefe experiments
when the gas had been breathed for a long time,
a confiderable diminution of it was obferved;
and the remainder extinguifhed flame and gave
a very flight diminution with nitrous gas. But
the great quantity of this remainder as well as
other phænomena, convinced me that though
the oiled filk was apparently air tight when

dry, under flight preſſure, yet during the action of reſpiration, the moiſt and warm gas expired, penetrated through it, whilſt common air entered through the wetted ſurface.

To aſcertain accurately, the changes effected in nitrous oxide by reſpiration, I was obliged to make uſe of the large mercurial airholder mentioned in Reſearch I. of the capacity of 200 cubic inches. The upper cylinder of it was accurately balanced ſo as to be conſtantly under the preſſure of the atmoſphere. To an aperture in it, a ſtop cock having a very large orifice was adapted, curved and flattened at its upper extremity, ſo as to form an air-tight mouth-piece.

By accurately cloſing the noſe, and bringing the lips tight on the mouth-piece, after a few trials I was able to breathe oxygene or common air in this machine for two minutes or two minutes and half, without any other uneaſy feeling than that produced by the inclination of the neck and cheſt towards the cylinder. The power of uniformly exhauſting the lungs and

,fauces to the fame extent, I did not acquire till after many experiments. At laft, by preferving exactly the fame pofture after exhauftion of the lungs before the infpiration of the gas to be experimented upon, and during its compleat expiration, I found that I could always retain nearly the fame quantity of gas in the bronchial veffels and fauces; the difference in the volume expired at different times, never amounting to a cubic inch and half.

By connecting the conducting pipe of the mercurial airholder, during the refpiration of the gas, with a fmall trough of mercury by means of a curved tube, it became a perfect and excellent breathing machine. For by exerting a certain preffure on the airholding cylinder, it was eafy to throw a quantity of gas after every infpiration or expiration, into tubes filled with mercury ftanding in the trough. In thefe tubes it could be accurately analifed, and thus the changes taking place at different periods of the procefs afcertained.

Whenever I breathed pure nitrous oxide in the mercurial airholder, after a compleat voluntary ex-

·hauſtion of my lungs, the pleaſurable delirium waſ
·very rapidly produced, and being obliged to ſtoop
on the cylinder, the determination of blood· to
my head from the increaſed arterial action in·
leſs than a minute became ſo great, as often to
deprive me of voluntary power over the muſcles
of the mouth. Hence, I could never rely on
the accuracy of any experiment, in which the
gas had been reſpired for more than three
·quarters of a minute.

I was able to reſpire the gas with great accuracy for more than half a minute; it at firſt,·
rather increaſing than diminiſhing the power of
volition; but even in this ſhort time, very ſtrong
ſenſations were always produced, with ſenſe of
fulneſs about the head, ſomewhat alarming; a
feeling which hardly ever occurs to me when the
gas is breathed in the natural poſture.

In all the numerous experiments that I made
on the reſpiration of nitrous oxide in this way,
a very conſiderable diminution of gas always
took place; and the diminution was generally
apparently greater to the eye during the firſt
four or five inſpirations,

The refidual gas of an experiment was always examined in the following manner. After being transferred through mercury into a graduated cylinder, a fmall quantity of concentrated folution of cauftic potafh was introduced to it, and fuffered to remain in contact with it for fome hours; the diminution was then noted, and the quantity of gas abforbed by the potafh, judged to be carbonic acid. To the remainder, twice its bulk of pure water was admitted. After agitation and reft for four or five hours, the abforption by this was noticed, and the gas abforbed confidered as nitrous oxide. The refidual unabforbable gas was mingled over water with twice its bulk of nitrous gas; and by this means, its compofition, whether it confifted wholly of nitrogene, or of nitrogene mingled with fmall quantities of oxygene, afcertained.

From a number of experiments made at different times, on the refpiration of nitrous oxide, I felect the following as the moft accurate.

E. 1. At temperature 54°, I breathed 102 cubic inches of nitrous oxide, which contained near $\frac{1}{50}$ common air, for about half a minute, feven infpirations and feven expirations being made. After every expiration, an evident diminution of gas was perceived ; and when the laft full expiration was made, it filled a fpace equal to 62 cubic inches.

Thefe 62 cubic inches analifed, were found to confift of

Carbonic acid ..	3,2
Nitrous oxide ..	29,0
Oxygene	4,1
Nitrogene	25,7
	62,0

Hence, accounting for the two cubic inches of common air previoufly mingled with the nitrous oxide, 71 cubic inches had difappeared in this experiment.

In the laft refpirations, the quantity of gas was fo much diminifhed, as to prevent the full expanfion of the lungs ; and hence the apparent diminution was very much less after the first four inspirations.

E. 2. At temperature 47°, I breathed 182 cubic inches of nitrous oxide, mingled with $2\frac{1}{2}$ cubic inches of atmospheric air, which previously existed in the airholder, for near 40 seconds; having in this time made 8 respirations. The diminution after the first full inspiration, appeared to a by-stander nearly uniform. When the last compleat expiration was made, the gas filled a space equal to 128 cubic inches, the common temperature being restored. These 128 cubic inches analised, were found to consist of

Carbonic acid	....	5,25
Nitrous oxide	....	88,75
Oxygene	.....	5,00
Nitrogene	......	29,00

Consequently, in this experiment, 93,25 cubic inches of nitrous oxide had disappeared.

In each of these experiments, the cylinder was covered with condensed watry vapor exactly in the same manner as if common air had been breathed in it. It ought to be obferved that, E. 1. was made in the morning, four hours and half after a moderate breakfaft; whereas, E. 2. was made but an hour and quarter after a plentiful dinner; at which near three-fourths of a pint of table-beer had been drank.

From thefe experiments we learn, that nitrous oxide is rapidly abforbed by the venous blood, through the moift coats of the pulmonary veins: But as after a compleat voluntary exhauftion of the lungs, much refidual air muft remain in the bronchial veffels and fauces, as appears from their incapability of compleatly collapfing, it is evident that the gas expired after every infpiration of nitrous oxide muft be mingled with different quantities of the refidual gas of the lungs;* whilft after a complete expiration, much of the unabforbed nitrous oxide muft remain as refidual gas in the lungs. Now when a complete expi-

---

* By lungs, I mean in this place, all the internal organs of refpiration.

ration is made after the breathing of atmofpheric air, it is evident that the refidual gas of the lungs confifts of nitrogene,* mingled with fmall portions of oxygene and carbonic acid. And thefe are the only products found after the refpiration of nitrous oxide.

To afcertain whether thefe products were partially produced, during the procefs of refpiration, as I was inclined to believe from the experiments in the laft fection, or whether they were wholly the refidual gafes of the lungs, I found extremely difficult.

I at firft thought of breathing nitrous oxide immediately after my lungs had been filled with oxygene; and to compare the products remaining after the full expiration, with thofe produced after a full expiration of pure oxygene; but on the fuppofition that oxygene and nitrous oxide, when applied together to the venous blood, muft effect changes in it different from

---

* Becaufe thefe products are formed during the refpiration of common air.

either of them feparately, the idea was relin-quifhed.

I attempted to infpire nitrous oxide, after hav-ing made two infpirations and a complete exph ration of hydrogene; but in this experiment the effects of the hydrogene were fo debilitating, and the confequent ftimulation by the nitrous oxide fo great, as to deprive me of fenfe. After the firft three infpirations, I loft all power of ftanding, and fell on my back, carrying in my lips the mouth-piece feparated from the cylinder, to the great alarm of Mr. Patrick Dwyer, who was noting the periods of infpiration.

Though experiments on fucceffive infpira-tions of pure nitrous oxide might go far to determine whether or no any nitrogene, car-bonic acid and oxygene were products of refpiration, yet I diftinctly faw that it was impoffible in this way to afcertain their quantities, fuppofing them produced, un-lefs I could firft determine the capacity of my lungs; and the different proportions of the gafes remaining in the bronchial veffels after a compleat expiration, when atmofpheric air had been refpired.

In fome experiments (that I made on the ref-piration of hydrogene, with a view to determine whether carbonic acid was *produced* by the combination of carbon loofely combined in the venous blood, with the oxygene refpired, or whether it was fimply *given out* as excrementitious by this blood) I found, without however being able to folve the problem I had propofed to myfelf, that in the refpiration of pure hydrogene, little or no alteration of volume took place; and that the refidual gas was mingled with fome nitrogene, and a little oxygene and carbonic acid.

From the comparifon of thefe facts with thofe noticed in the laft fection and in R. III. Div. I. there was every reafon to fuppofe that hydrogene was not abforbed or altered when refpired; but only mingled with the refidual gafes of the lungs. Hence, by making a full expiration of atmofpheric air, and afterwards taking fix or feven refpirations of hydrogene in the mercurial

airholder, and then making a compleat expiration, I conjectured that the residual gas and
the hydrogene would be fo mingled, as that
nearly the fame proportions fhould remain in the
bronchial veffels, as in the airholder. By afcertaining thefe proportions and calculating from
them, I hoped to be able to afcertain with
tolerable exactnefs, the capacity of my fauces
and bronchia, as well as the compofition of the
gas remaining in them, after a complete expiration of common air.

IV. *Refpiration of Hydrogene.*

The hydrogene that I employed, was procured from the decompofition of water by means
of clean iron filings and diluted fulphuric and
muriatic acids. It was breathed in the fame
manner as nitrous oxide, in the large mercurial
airholder.

After a compleat voluntary exhauftion of
my lungs in the ufual pofture, I found great
difficulty in breathing hydrogene for fo long

as half a minute, fo as to make a compleat expiration of it. It produced uneafy feelings in the cheft, momentary lofs of mufcular pow-er, and fometimes a tranfient giddinefs.

In fome of the experiments that I made; on account of the giddinefs, the refults were ren-dered inconclufive, by my removing my mouth from the mouth-piece after expiration, before the affiftant could turn the ftopcock.

The purity of the hydrogene was afcertained immediately before the experiment by the teft of nitrous gas, and by detonation with oxygene or atmofpheric air; generally 12 meafures of atmofpheric air were fired with 4 of the hydro-gene, and if the diminution was to ten or a little more, the gas was judged to be pure.

After the experiment, when the compleat expiration had been made and the common temperature reftored; the volume of the gas was noticed, and then a fmall quantity of it thrown into the mercurial apparatus by means of the conducting tube, to be examined. The carbonic acid was feparated by from it by means

B b

of folution of potafh or ftrontian ; the quantity of oxygene it contained, was afcertained by means of nitrous gas of known compofition ; the fuperabundant nitrous, gas was abforbed by folution of muriate of iron ; and the proportions of hydrogene and nitrogene in the remaining gas, difcovered by inflammation with atmofpheric air, or oxygene in the detonating tube by the electric fpark.

a. The two following experiments made upon quantities of hydrogene, equal to thofe of the nitrous oxide refpired in the experiments in the laft fection, are given as the moft accurate of five.

E. 1. I refpired at 59° 102 cubic inches of hydrogene apparently pure, for rather lefs than half a minute, making, in this time feven, quick refpirations.

After the complete expiration, when the common temperature was reftored, the gas occupied a fpace equal to 103 cubic inches, nearly. Thefe analifed were found to confift of

Carbonic acid ..	4,0
Oxygene	3,7
Nitrogene	17,3
Hydrogene	78,0
	103,0

Now as in this experiment, the gas was increafed in bulk only a cubic inch ; fuppofing that after: the compleat expiration the gas in the lungs; bronchia and fauces was of nearly fimilar com-¦ pofition with that in the airholder, and that no. hydrogene had been abforbed by the blood, it would follow that 24 cubic inches of hydrogene remained in the internal organs of refpiration, and confequently, by the rule of proportion, about 7,8 of the mixed refidual gas of the common air. And then the whole quantity of refidual gas of the lungs, fuppofing the temperature 59°, would have been 31,8 cubic inches ; but as its temperature was nearly that of the internal parts of the body, 98°, it muft have filled a greater fpace; calculating from the experiments.

of Guyton and Vérnois,* about 37,5† cubic inches.

From the increafe of volume, it would appear that a minute quantity of gas had been generated during the refpiration, and this was, as we fhall fee hereafter, moft probably carbonic acid.§ Likewife there is reafon to fuppofe, that a little of the refidual oxygene muft have been abforbed. Making allowances for thofe circum-ftances, it would follow, that the 37,5 cubic inches of gas remaining in my lungs, after a compleat expiration of atmofpheric air at animal heat 98°, equal to 31,8 cubic inches at 59°, were compofed of.

> Nitrogene 21,9
> Carbonic acid .. 4,9
> Oxygene 5,0
> ─────
> 31,8

* Annales de Chimie, vol. 1, page 279.

† This is only an imperfect approximation; the ratio of the increafe of expanfibility of gafes to the increafe of temperature, has not yet been afcertained. It is probable that the expanfibility of gafes is altered by their mixture.

§ For there is no reafon to fuppofe the production of nitrogene.

E. 2. - I refpired for near a half a minute in
the mercurial airholder at 61°, 182, cubic inches of hydrogene; having made during this
time, fix long infpirations. After the laft expiration, the gas filled a fpace nearly equal to
184 cubic inches, and analifed, was, found to
confift of

       Carbonic acid   4,8
       Oxygene '   4,6
       Nitrogene 21,0
       Hydrogene 153,6
                  ———
                184.

Now in this experiment, reafoning in the fame
manner as before, 28,4 cubic inches of hydrogene muft have remained in the lungs, and
likewife 5,5 of the atmofpheric refidual gas:
Confequently, the whole refidual gas was nearly
equal to 34 cubic inches at 61o, which at 98°
would become about 40,4 cubic inches. · And
reafoning as before, it would appear from this
experiment, that the quantity of gas remaining
in my lungs after a compleat voluntary refpira-

tion, equalled at 98, about 40 cubic inches,
and at 61°, 34 nearly : making the neceffary
corrections; that after common air had been
breathed, thefe 34 cubic inches confifted of

    Carbonic acid 4,1
    Oxygene 5,5
    Nitrogene 24,4

b. It would have been poffible to prove the truth
of the poftulate on which the experiments were
founded, by refpiring common air or oxygene
after the compleat expiration of the hydrogene,
for the fame time as the hydrogene was refpired and in equal quantities.

For if portions of hydrogene were found in
the airholder equal to thofe of the refidual gafes
in the two experiments, it would prove that a.
uniform mixture of refidual gas with the gas
infpired, was produced by the refpiration. That
this mixture muft have taken place, appeared,
however, fo evident from analogous facts, that
I judged the experimental proof unneceffary.

Indeed, as moft gafes, though of different fpecific gravities, when brought in contact with each

other, affume fome fort of union, it is more
than probable, that gas infpired into the lungs,
from being placed in contact with the residual
gas on fuch an extenfive furface, muft inftantly
mingle with it. Hence, poffibly one deep infpiration and compleat expiration of the whole
of a quantity of hydrogene, will be fufficient to
determine the capacity of the lungs after compleat voluntary exhauftion, and the nature of
the refidual air.

That two infpirations are fufficient, appears
probable from the following experiment.

E. 3. After a compleat voluntary expiration
of common air, I made two deep infpirations of
141 cubic inches of hydrogene. After the
compleat expiration, they filled a fpace equal
to rather more than 142 cubic inches, and
analifed, were found to confift of

> Carbonic acid 3,1
> Oxygene 4,5
> Nitrogene 18,8
> Hydrogene 115,6
> ——————
> 142.

Now calculating on the exhaufted capacity of my lungs from this experiment, fuppofing uniform mixture, they would contain after expiration of common air, about 30,7 cubic inches at 58°, equal to 36 at 98°, compofed of about

Nitrogene 20,9
Oxygene 5,8
Carbonic acid .. 4,0

30,7

One fhould fuppofe a priori that in this experiment much lefs of the refidual oxygene of the lungs muft have been abforbed, than in Expts. 1 and 2 ; yet there is no very marked difference in the portions evolved. That a tolerably accurate mixture took place, appears from the quantity of nitrogene. The fmaller quantity of carbonic acid is an evidence in favour of its evolution from the venous blood.

c. It is reafonable to fuppofe that the preffure upon the refidual gas of the exhaufted lungs, muft be nearly equal to that of the atmofphere. But as aqueous vapour is perpetually given out by the exhalents, and perhaps evolved from the moiſt coats of the pulmonary veſſels, it is likely that the reſidual gas is not only fully ſaturated with moiſture at 98°, but likewiſe impregnated with uncombined vapor; and hence its volume enlarged beyond the increment of expanſion of temperature.

, Conſidering all theſe circumſtances, and calculating from the mean of the three experiments on the compoſition of the reſidual gas, I concluded,

1ſt. That the exhauſted capacity of my lungs was equal to about 41 cubic inches.

, 2dly. That the gas contained in my bronchial veſſels and fauces, after a compleat reſpiration of atmoſpheric air, was equal to about 32 cubic inches, its temperature being reduced to 55°.

3dly. That theſe 32 cubic inches were compoſed of about

Nitrogene ..	23,0
Carbonic acid ..	4,1
Oxygene . ..	4,9

d. In many experiments made in the mercurial airholder on the capacity of my lungs under different circumſtances, I found that I threw out of my lungs by a full forced expiration at temperatures from 58° to 62°

cub. in. cub. in.

After a full voluntary inſpiration, from 189 to 191

After a natural inſpiration, from .. 78 to 79

After a natural expiration, from .. 67 to 68

So that making the corrections for temperature, it would appear, that my lungs in a ſtate of voluntary inſpiration, contained about 254 cubic inches; in a ſtate of natural inſpiration about 135 ; in a ſtate of natural expiration, about 118 ; and in a ſtate of forced expiration 41.*

As the exhauſted capacity as well as impleted capacity of the internal organs of reſpiration muſt be different in different individuals, according as the forms and ſize of their thorax,

---

* This capacity is moſt probably below the medium, my cheſt is narrow, meaſuring in circumference, but 29 inches, and my neck rather long and ſlender.

fauces, and bronchia are different, it would be almoſt uſeleſs to endeavour to aſcertain a ſtandard capacity. It is however probable, that a ratio exiſts between the quantities of air inſpired in the natural and forced inſpiration; thoſe expired in the natural and forced expiration, and the whole capacity of the lungs. If this ratio were aſcertained, a ſingle experiment on the natural inſpiration and expiration of common air, would enable us to aſcertain the quantity of reſidual gas in the lungs of any individual after a compleat forced expiration.*

V. *Additional obſervations and experiments on the Reſpiration of Nitrous Oxide.*

a. Having thus aſcertained the capacity of my lungs, and the compoſition of the reſidual gas of expiration, I proceeded to reaſon concerning

---

* Dr. Goodwyn in his excellent work on the connexion of life with reſpiration, has detailed ſome experiments on the capacity of the lungs after natural expiration. He makes the medium capacity about 109 cubic inches, which agrees very well with my eſtimation.—page 27.

the experiments in section III, on the respiration. of nitrous oxide.

In Exp. I. nearly 100 cubic inches of nitrous oxide, making the corrections on account of the common air, were respired for half a minute. In this time, they were reduced to 62 cubic inches, which consisted of 3,2 carbonic acid, 29 nitrous oxide, 4,1 oxygene, and 25,7 nitrogene.

But, as appears from the last section, there existed in the lungs before the inspiration of the nitrous oxide, about 32 cubic inches of gas, consisting of 23 nitrogene, 4,1 carbonic acid, and 4,9 oxygene, temperature being reduced to 59°. This gas must have been perfectly mingled with the nitrous oxide during the experiment; and consequently, the residual gas in the lungs after the experiment, was of the same composition as that in the airholder.

Supposing it as before, to be about 32 cubic inches: from the rule of proportion, they will be composed of

Nitrous oxide .. 14,7

Nitrogene 13,3

Carbonic acid 1,9

Oxygene 47 2,1

And the whole quantity of gas in the lungs and the airholder, supposing the temperature 59°, will equal 94 cubic inches, which are composed of

Nitrous oxide ... 43,7

Nitrogene 39,0

Carbonic acid 5, 2

Oxygene 6, 1

———

94

But before the experiment, the gas in the lungs and airholder equalled 134 cubic inches, and thefe, reckoning for the common air, were compofed of

Nitrous oxide .. 100

Nitrogene, :... 24,3

Carbonic acid ... 4,1

Oxygene 5,6

Hence, it appears, that 56,3 cubic inches of nitrous oxide were abforbed in this experiment, and 13,7 of nitrogene produced, either by evolution from the blood, or decompofition of the nitrous oxide. The quantities of carbonic acid and oxygene approach fo near to thofe exifting after the refpiration of hydrogene, that there is every reafon to believe that no portion of them was produced in confequence of the abforption, or decompofition of the nitrous oxide.

b. In Exp. 2, calculating in the fame manner, before the firft infpiration, a quantity of gas equal to 216,5 cubic inches at 47°, exifted in the lungs and airholder, and thefe 216,5 cubic inches were compofed of

Nitrous oxide,	. . . .	182,0
Nitrogene	. . . . .	24,9
Carbonic acid	. . . .	4,1
Oxygene	. . . . . .	5,5
		216,5

After the compleat expiration, 160 cubic in- ches remained in the lungs and airholder, which was compofed of

Nitrous oxide	. .	110,6
Nitrogene	. . . . . .	36,3
Carbonic acid	. . . .	6,8
Oxygene	. . . .	6,3

Hence, it appears, that 71,4 cubic inches of nitrous oxide were abforbed in this experiment, and about 12, of nitrogene produced. The quantity of carbonic acid and oxygene is rather greater than that which exifted in the experiments, on hydrogene.

c. From, thefe eftimations, I learned that a fmall quantity of nitrogene was produced during the abforption of nitrous oxide in refpiration. It remained to determine, whether this nitrogene owed its production to evolution from the blood, or to the decompofition of a portion of the nitrous oxide.

Analogical evidences were not in favour of the hypothefis, of decompofition. It was difficult to fuppofe that a body requiring the temperature of ignition for its decompofition by the moft inflammable bodies, fhould be partially abforbed, and partially decompounded at 98° by a fluid apparently poffeffed of uniform, attractions.

It was more eafy to believe, that from the immenfe quantity of nitrogene taken into the blood, in nitrous oxide; the fyftem foon became.

overcharged with this principle, which not being wholly expended in new, combinations during living action, was liberated in the aëriform ſtate by the exhalents, or through the moiſt coats of the veins.

Now if the laſt rationale were true, it would follow, that the quantity of nitrogene produced in reſpiration, ought to be increaſed in proportion as a greater quantity of nitrous oxide entered into combination with the blood.

d. To aſcertain whether this was the caſe, I made after full voluntary exhauſtion of my lungs, one full voluntary inſpiration and expiration of 108 cubic inches of nitrous oxide. After this, it filled a ſpace nearly equal to 99 cubic inches. The quantities of carbonic acid and oxygene in theſe were not determined; but by the teſt of abſorption by water, they appeared to contain only 18 nitrogene; which is very little more than ſhould have been given from the reſidual gas of the lungs.

In a ſecond experiment, I made two reſpirations of 108 cubic inches of nitrous oi de nearly pure. The diminution was to 95. On analyfing thefe 95, I found to my great furprife, that they contained only 17 nitrogene. Hence, I could not but fufpect fome fource of error in the procefs.

I now introduced into a ftrong new filk bag, the fides of which were in perfect contact, about 8 quarts of nitrous oxide. From the mode of introduction, this nitrous oxide muft have been mingled with a little common air, not however, fufficient to difturb the refults.

I then adapted a cork cemented to a long curved tube to my right noftril ; the tube was made to communicate with the water apparatus; and the left noftril being accurately clofed, and the mouth-piece of the filk bag tightly adapted to the lips, I made a full expiration of the common air of my lungs, infpired nitrous oxide from the bag, and by carefully clofing the mouth-piece with my tongue, expired it through the curved tube into the water apparatus. In this way, I made nine refpirations of nitrous oxide. The expired gas of the firft refpiration was not preserved ; but part of the gas of the second, third, fifth, seventh and ninth, were caught in feperate graduated cylinders. The second, analifed by abforption, confifted of about 29 abforbable gas, which muft have been chiefly nitrous oxide ; and 17 unabforbable gas, which muft have been chiefly nitrogene; and the third of 22 abforbable gas, and 8 unabforbable. The fifth was compofed of 27 to 6 ; the feventh of 23 to 7, and the ninth of 26 to 11.

e. Though the refults of thefe experiments were not fo conclufive as could be wifhed ; yet, comparing them with thofe of the experiments in fection III. it feemed reafonable to conclude, that the production of nitrogene was increafed, in proportion as the blood became more fully impregnated with nitrous oxide.

From this conclufion, compared with the phæ-nomenon noticed in fection 2, and in Div. I. fection 4, I am induced to believe that the production of nitrogene during the refpiration of nitrous oxide, is not owing to the decompofi-tion of part of the nitrous oxide, in the aëriform ſtate immediately by the attraction of the red particles of venous blood for its oxygene; but that it is rather owing to a new arrangement produced in the principles of the impregnated blood, during circulation; from which, becoming ſuperſaturated with nitrogene, it gives it out through the moiſt coats of the veſſels.

For if any portion of nitrous oxide were de-compoſed immediately by the red particles of the blood, one ſhould conjecture, that the quantity of nitrogene produced, ought to be greater during the firſt inſpirations, before theſe particles became fully combined with condenſed oxygene. If, on the contrary, the whole of the nitrogene and oxygene of the nitrous oxide were both combined with the blood, and carried through the pulmonary veins and left chamber of the heart to the arteries; then, ſuppoſing the oxygene chiefly expended in living action, whilſt the nitrogene was only partially conſumed in new combina-tions, it would follow, that the venous blood of animals made to breathe nitrous oxide, hyper-ſaturated with nitrogene, muſt be different from

common venous blood ; and this we have reafon to believe from the phænomena' in Div. I.
fection 4, is actually the cafe.

f. Befides the nitrogene generated during
the refpiration of nitrous oxide, we have noticed
the evolution of other products, carbonic acid,*
and water.

Now as nearly equal quantities of carbonic
acid are produced, whether hydrogene or nitrous oxide is refpired, provided the procefs is
carried on for the fame time ; there is every
reafon to believe, as we have faid before, that no
part of the carbonic acid produced, is generated
from the immediate decompofition of nitrous
oxide by carbon exifting in the blood.

Confequently, in thefe experiments, it muft
be either evolved from the venous blood ; or
formed, by the flow combination of the oxygene
of the refidual air of refpiration with the charcoal of the blood.

---

* The oxygene as we have before noticed, moft probably wholly exifted in the refidual gas.

But if it was produced by the decompofition of refidual atmofpheric air, it would follow, that its volume muft be much lefs than that of the oxygene of the refidual air, which had difappeared ; for fome of this oxygene muft have been *abforbed* by the blood, and during the converfion of oxygene into carbonic acid by charcoal, a flight diminution of volume is produced.

In the experiments, when nitrous oxide and hydrogene were refpired for about half a minute, the medium quantity of carbonic acid produced, was 5,6 cubic inches nearly.

Now we will affume, that the quantity of carbonic acid produced, is in the ratio of the oxygene diminifhed ; and there is every reafon to believe, that in the expiration of atmofpheric air, the expired air and the refidual air are nearly of the fame compofition.

Hence, no more carbonic acid can remain in the lungs or be produced from the refidual gas after the compleat expiration of common air, than that which can be generated from a volume of atmofpheric air equal to the refidual gas of the lungs.

The refidual gas of the lungs, after compleat expiration, equals at 55o, 32 cubic inches, and 32 cubic inches of common air contain 8.6 cubic inches of oxygene.

But in the experiments on the refpiration of hydrogene, not only 5.6 cubic inches of carbonic acid were produced, but more than 4 of refidual oxygene remained unabforbed.

Hence it appears impoffible that all the carbonic acid evolved from the lungs during the refpiration of nitrous oxide or hydrogene could have been produced by the combination of charcoal in the venous blood with refidual atmofpheric oxygene : there is confequently every reafon to believe that it is wholly or partially liberated from the venous blood through the moift coats of the veffels.

g. The water carried out of the lungs in folution by the expired gas of nitrous oxide, could neither have been wholly or partially formed by the decompofition of nitrous oxide. The

coats of the veffels in the lungs, and indeed in
the whole internal furface of the body, are
always covered with moifture, and the folution
of part of this moifture by the infpired heated
gas, and its depofition by the expired gas, are
fufficient caufes for the appearance of the
phænomenon.

There are no reafons for fuppofing that any
of the refidual atmofpheric oxygene is immediately combined with fixed or nafcent hydrogene; or hydrocarbonate, in the venous blood at
98°, by flow combuftion, and confequently
none for fuppofing that water is immediately
formed in refpiration.

The evolution of water from the veffels in
the lungs, is almoft certain from numerous
analogies.

b. As from the experiments in fection II. it
appeared that nitrous oxide was capable of being
combined with oxygenated blood, and vice verfa,
blood impregnated with nitrous oxide capable
of oxygenation ; I was curious to afcertain what
changes would be effected in nitrous oxide when

it was respired, mingled with atmospheric air or oxygene. For this purpose, without making a very delicate experiment, I breathed in the large mercurial airholder about 112
cubic inches of nitrous oxide, mingled with
44 of common air, for near half a minute, in
the usual mode. The gas, after expiration, filled a space nearly equal to 119. I did not
exactly ascertain the composition of the residual
gas; it supported flame rather better than common air, and after the nitrous oxide was absorbed, gave much less diminution with nitrous
gas than atmospheric air.

i. I breathed a mixture of four quarts of
nitrous oxide with three quarts of hydrogene,
in a dry silk bag, for near a minute; an evident diminution was produced; but on account
of the mode of experimenting it was impossible
to determine the quantity of nitrous oxide absorbed, or the exact nature of the products.
When a taper was introduced into a little of the
residual gas, it inflamed with a very feeble explosion. Now a mixture of 4 parts nitrous ox-

ide, and 3 hydrogene, detonates when inflamed with very great violence.

k. Nitrous oxide can be respired without danger by the human animal for a much longer time than that required for the death of the smaller quadrupeds in it.

I have breathed it two or three times in a confiderable ftate of purity, in a dry filk bag, for four minutes and quarter and four minutes and half: fome difeafed individuals have refpired it for upwards of five minutes.

In the infancy of my experiments, from general appearances, I thought that the proportion of nitrous oxide abforbed in refpiration was greater in the firft infpirations than the laft; but this I have fince found to be a miftake. In the laft refpirations the apparent abforption is indeed lefs; but this is on account of the increafed evolution of nitrogene from the blood. When nitrous oxide is refpired for a long time, the laft infpirations are always fuller and quicker than the firft; but the confumption by the fame individual is nearly in the ratio of the time of refpiration. Three quarts i. e. about 174 cubic inches, are consumed so as to be unfit for respiration, by an healthy individual with lungs of moderate capacity, in about a minute and quarter ; six quarts, or 348 cubic inches, last generally for two minutes and half or two minutes and three quarters ; eight quarts, or 464 cubic inches, for more than three minutes and half; and twelve, or 696 cubic inches, for nearly five.

The quantities of nitrous oxide absorbed by the same individual, will, as there is every reason to suppose, be different under different circumstances, and will probably be governed in some measure by the state of the health. It is reasonable to suppose, that the velocity of the circulation must have a considerable influence on the absorption of nitrous oxide ; probably in proportion as it is greater a larger quantity of gas will be consumed in equal times.

I am inclined from two or three experiments, to believe that nitrous oxide is absorbed more rapidly after hearty meals or during stimulation from wine or spirits, than at other times. As its abforption appears to depend on a fimple folution in the venous blood; probably diminution of temperature will increafe its capability of being abforbed.

l. The quantities of nitrous oxide abforbed by different individuals, will probably be governed in fome meafure by the fize of their lungs, and the furface of the blood veffels, all other circumftances being the fame.

From the obfervations that I have been able to make on the abforption of nitrous oxide, as compared with the capacity of the lungs, the range of the confumption of different individuals does not extend to more than a pint, or 30 cubic inches at the maximum dofe.

We may therefore conclude, that the medium confumption of nitrous oxide by the refpiration of different individuals, is not far from two cubic inches, or about a grain every fecond, or 120 cubic inches, or 60 grains every minute.

m. When nitrous oxide is breathed in tight filk bags, towards the end of the experiment as the internal furface becomes moift, as I have before mentioned, a certain quantity of common air penetrates through it and becomes mixed with the refidual gas, of the experiment; but this quantity is always too fmall to deftroy any of the effects of the nitrous oxide. The refidual gas of the common air, the nitrogene and carbonic acid produced in the procefs, and the refiduum of the admitted atmofpheric air, hardly ever amount after the experiment, to one half of the volume of the nitrous oxide abforbed. There is confequently, a perfect propriety in fucceffively infpiring and expiring the whole of a given quantity of nitrous oxide, till it is nearly confumed. In the refpiration of nitrous oxide as the gas is abforbed and not decompofed, little will be gained in effect, by perpetually infpiring and expiring new portions, whilft an immenfe quantity of gas will be idly wafted, and this circumftance, confidering the expence of the fubftance, is of importance,

VI. *On the respiration of Atmospheric Air.*

Having thus afcertained the abforption of nitrous oxide in refpiration, and the evolution of nitrogene and carbonic acid from the lungs during its abforption : confidering atmofpheric air as a compound in which principles identical with thofe in nitrous oxide exifted, though in different quantities and loofer combination, I was anxious to compare the changes effected in this gas by refpiration, with thofe produced in nitrous oxide and oxygene ; particularly as they are connected with the health and life of animals.

The ingenious experiments of Lavoifier and Goodwyn, prove the confumption of oxygene in refpiration, and the production of carbonic acid. From many experiments on the refpiration of common air, Dr. Prieftley fufpected that a certain portion of nitrogene, as well as oxygene, was abforbed by the venous blood.

b. In the following experiments on the refpiration of atmofpheric air in the mercurial airholder ; the compofition of the gas before infpiration and after expiration, was afcertained in
the following manner.

Forty meafures.of it were agitated over mercury
in folution of cauftic potafh,and fuffered to remain,
in contact with it for two or three hours. The
diminution was noted, and the gas abforbed
judged to be carbonic acid. Twenty meafures
of the gas, freed from carbonic acid, were mingled with thirty of nitrous gas, in a tube of ,5
inches diameter ; they were not agitated,* but
fuffered to reft for an hour or an hour and half,
when the volume occupied by them was noticed;
and 50 — m the volume occupied, divided
by 3 confidered as the oxygene x, and 20 — x
confidered as the nitrogene.

* When they are agitated, a greater proportion of nitrous
gas is abforbed, condenfed in the nitric acid by the water ;
and to find the oxygene,

$$x = \frac{50 - m}{3,4} \text{ or } \frac{50 - m}{3,5}$$

c. To, afcertain the changes effected in at-mofpheric air by fingle infpirations, . .

I made, after a compleat voluntary exhauftion of my lungs, at temperature 61°, one infpiration and expiration of 141 cubic. inches of atmofpheric air. After expiration, they filled a fpace equal to 139 cubic inches nearly. Thefe 139 cubic inches analifed were found to confift of

> Nitrogene 101
> Oxygene. 32
> Carbonic acid .. 6

The 141 cubic, inches before infpiration, were compofed of 103 nitrogene, 1 carbonic acid and 37 oxygene. The time taken to per-form the infpiration and full-expiration, was nearly a quarter of a minute.

I repeated this experiment feven or eight times, and the quantity of oxygene abforbed was generally from 5 to 6 cubic inches, the carbonic acid formed from 5 to 5,5, and the quantity of nitrogene apparently diminifhed by from 1 to 3 cubic inches. --

E. 2. I made, after a voluntary expiration of common air, one infpiration and full expiration of 100 cubic inches of atmofpheric air. It was diminifhed nearly to 98¾ or 99 cubic inches, and analifed, was found to confift of

Nitrogene .. 71,7
Oxygene 22,5
Carbonic acid .. 4,5

This experiment I likewife repeated four or five times, with very little difference of refult, and there always feemed to be a fmall diminution of nitrogene. I made no corrections on account of the refidual air of the lungs in thefe proceffes, becaufe there was every reafon to fuppofe that it was always of fimilar compofition.

c. Before I could afcertain whether fimilar changes were effected in atmofpheric air, by natural infpirations as by forced ones, I was obliged to practife refpiration in the mercurial airholder, by fuffering the conducting tube to communicate with the atmofphere till I had attained the power of breathing in it naturally, without labor or attention ; I then found by a

IX. *Obſervations on the reſpiration of Nitrous Oxide.*

The experiments in the firſt Diviſion of this Reſearch, prove that nitrous oxide when reſpired by animals, produces peculiar changes in their blood and in their organs, firſt conneſted with increaſed living aſtion; but terminating in death.

From the experiments in this Diviſion, it appears, that nitrous oxide is rapidly abſorbed by the circulating venous blood, and of courſe its condenſed oxygene and nitrogene diſtributed in the blood over the whole of the ſyſtem.

Concerning the changes effeſted in the principles of the impregnated blood during circulation and its aſtion upon the nervous and muſcular fibre ; it is uſeleſs to reaſon in the preſent ſtate of our knowledge.

It would be eaſy to form theories referring the aſtion of blood impregnated with nitrous oxide, to its power of ſupplying the nervous and muſcular fibre with ſuch proportions of condenfed nitrogene, oxygene and light or etherial fluid, as enabled them more rapidly to pafs through thofe changes which conftitute their life : but fuch theories would be only collections of terms derived from known phænomena and applied by loofe analogies of language to un-known things.

We are unacquainted with the compofition of dead organifed matter ; and new inftruments of experiment and new modes of refearch muft be found, before we can afcertain even our capabilities of difcovering the laws of life.

RESEARCH IV.

RELATING TO THE

EFFECTS PRODUCED BY THE RESPIRATION

OF

NITROUS OXIDE

UPON DIFFERENT

INDIVIDUALS.

RESEARCH IV.

THE EFFECTS

PRODUCED BY THE

RESPIRATION of NITROUS OXIDE.

DIVISION I.

HISTORY of the DISCOVERY.—Effects produced by the RESPIRATION of different GASES.

A SHORT time after I began the study of Chemiftry, in March 1798, my attention was directed to the dephlogifticated nitrous gas of Prieftley, by Dr. Mitchill's Theory of Contagion.*

The fallacy of this Theory was foon demon-ftrated, by a few coarfe experiments made on fmall quantities of the gas procured from zinc

* Dr. Mitchill attempted to prove from fome phænome-na connected with contagious difeafes, that dephlogifticated nitrous gas which he called oxide of fepton, was the prin-ciple of contagion, and capable of producing the moft terri-ble effects when refpired by animals in the minuteft quantities or even when applied to the fkin or mufcular fibre.

and diluted nitrous acid. Wounds were expofed to its action, the bodies of animals were
immerfed in it without injury; and I breathed
it mingled in fmall quantities with common air,
without remarkable effects. An inability to
procure it in fufficient quantities, prevented me
at this time, from purfuing the experiments to
any greater extent. I communicated an account of them to Dr. Beddoes.

In 1799, my fituation in the Medical Pneumatic Inftitution, made it my duty to inveftigate the phyfiological effects of the aëriform
fluids, the properties of which prefented a
chance of ufeful agency. At this period I re-
commenced the inveftigation.

A confiderable time elapfed before I was
able to procure the gas in a ftate of purity, and
my firft experiments were made on the mixtures
of nitrous oxide, nitrogene and nitrous gas,
which are produced during metallic folutions.

In the beginning of March, I prepared a large
quantity of impure nitrous oxide from the nitrous folution of zinc. Of this I often breathed

the quantities of a quart and two quarts generally mingled with more than equal parts of
. oxygene or common air. In the moſt decifive
of thoſe trials, its effects appeared to be depreſ-
fing, and I imagined that it produced a tendency to fainting : the pulſe was certainly
rendered flower under its operation.

At this time, Mr. Southey refpired it in an
highly diluted ſtate ; it occafioned a flight degree
of giddineſs, and confiderably diminiſhed the
quickneſs of his pulſe.

Mr. C. Coates likewiſe refpired it highly diluted, with fimilar effects.

In April, I obtained nitrous oxide in a ſtate
of purity, and afcertained many of its chemical
properties. Reflections upon thefe properties
and upon the former trials, made me refolve to
endeavour to infpire it in its pure form, for I
faw no other way in which its refpirability, or
powers could be determined.*

* I did not attempt to experiment upon animals, becaufe
they die nearly in equal times in non-refpirable gafes, and
gafes incapable of fupporting life, and poffeſſed of an
action on the venous blood.

I was aware of the danger of this experiment. It certainly would never have been made if the hypothefis of Dr. Mitchill had in the leaft influenced my mind. — I thought that the effects might be poffibly depreffing and painful, but there were many reafons which induced me to believe that a fingle infpiration of a gas apparently poffeffing no immediate action on the irritable fibre, could neither deftroy or materially injure the powers of life.

On April 11th, I made the firft infpiration of pure nitrous oxide; it paffed through the bronchia without ftimulating the glottis, and produced no uneafy feeling in the lungs.

The refult of this experiment, proved that the gas was refpirable, and induced me to believe that a farther trial of its effects might be made without danger.

On April 16th, Dr. Kinglake being accidentally prefent, I breathed three quarts of nitrous oxide from and into a filk bag for more than half a minute, without previoufly clofing my nofe or exhaufting my lungs.

The firſt inſpirations occaſióned a ſlight degree of giddineſs. This was ſucceeded by an uncommon ſenſe of fulneſs of the head, accompanied with loſs of diſtinct ſenſation and voluntary power, a feeling analogous to that produced in the firſt ſtage of intoxication; but unattended by pleaſurable ſenſation. Dr. Kinglake, who felt my pulſe, informed me that it was rendered quicker and fuller.

This trial did not ſatisfy me with regard to its powers; comparing it with the former ones I was unable to determine whether the operation was ſtimulant or depreſſing:

I communicated the reſult to Dr. Beddoes, and on April the 17th, he was preſent, when the following experiment was made.

Having previouſly cloſed my noſtrils and exhauſted my lungs, I breathed four quarts of nitrous oxide from and into a ſilk bag. The firſt feelings were ſimilar to thoſe produced in the laſt experiment; but in leſs than half a minute, the reſpiration being continued, they diminiſhed gradually, and were ſucceeded by a fenfation analogous to gentle preſſure on all the muſcles, attended by an highly pleaſurable thrilling, particularly in the cheſt and the extremities. "The objects around me became dazzling and my hearing more acute. Towards the laſt inſpirations, the thrilling increaſed, the ſenſe of muſcular power became greater, and at laſt an irreſiſtible propenſity to action was indulged in; I recollect but indiſtinctly what followed; I know that my motions were various and violent.

Theſe effects very ſoon ceaſed after reſpiration. In ten minutes, I had recovered my natural ſtate of mind. The thrilling in the extremities, continued longer than the other ſenſations.*

This experiment was made in the morning; no languor or exhauſtion was conſequent, my feelings throughout the day were as uſual, and I paſſed the night in undiſturbed repoſe.

* Dr. Beddoes has given ſome account of this experiment, in his Notice of ſome obſervations made at the Medical Pneumatic Inſtitution. It was noticed in Mr. Nicholſon's Phil. Journal for May 1799.

The next morning the recollections of the effects of the gas were very indiftinct, and had not remarks written immediately after the experiment recalled them to my mind, I fhould have even doubted of their reality. I was willing indeed to attribute fome of the ftrong emotion to the enthufiafm, which I fuppofed muft have been neceffarily connected with the perception of agreeable feelings, when I was prepared to experience painful fenfations. Two experiments however, made in the courfe of this day, with fceptifm, convinced me that the effects were folely owing to the fpecific operation of the gas.

In each of them I breathed five quarts of nitrous oxide for rather a longer time than before. The fenfations produced were fimilar, perhaps not quite fo pleafurable ; the mufcular motions were much lefs violent.

Having thus afcertained the powers of the gas, I made many experiments to afcertain the length of time for which it might be breathed with fafety, it's effects on the pulfe, and its

general effects on the health; when often refpired.

I found that I could breathe nine quarts of
nitrous oxide, for three minutes, and twelve
quarts for rather more than four. I could
never breathe it in any quantity, fo long as five
minutes. Whenever its operation was carried
to the higheft extent, the pleafurable thrilling
at its height about the middle of the experiment,
gradually diminifhed, the fenfe of preffure on
the mufcles was loft; impreffions ceafed to be
perceived; vivid ideas paffed rapidly through
the mind, and voluntary power was altogether
deftroyed, fo that the mouth-piece generally
dropt from my unclofed lips.

Whenever the gas was in a high ftate of purity, it tafted diftinctly fweet to the tongue and
palate, and had an agreeable odor. I often
thought that it produced a feeling fomewhat
analogous to tafte, in its application to my lungs.
In one or two experiments, I perceived a diftinct
fenfe of warmth in my cheft.

I never felt from it any thing like oppreffive

refpiration : my infpirations became deep in proportion as I breathed it longer; but this phænomenon arofe from increafed energy of the mufcles of refpiration, and from a defire of increafing the pleafurable feelings.

Generally when I breathed from fix to feven quarts, mufcular motions were produced to a certain extent ; fometimes I manifefted my pleafure by ftamping or laughing only ; at other times, by dancing round the room and vociferating.

After the refpiration of fmall dofes, the exhilaration generally lafted for five or fix minutes only. In one or two experiments when ten quarts had been breathed for near four minutes, an exhilaration and a fenfe of flight intoxication lafted for two or three hours.

On May 3d, To afcertain whether the gas would accelerate or retard the progrefs of fleep, I breathed at about 8 o'clock in the evening, 25 quarts of nitrous oxide, in quantities of fix at a time, allowing but fhort intervals between each dofe. The feelings were much lefs pleafurable than ufual, and during the confumption of the two, laft doſes, almoſt indifferent ; indeed the gas was breathed rather too foon after its production and contained, ſome ſuſpended acid vapour which ſtimulated the lungs ſo as to induce coughing.

After the experiments, for the firſt time I was ſomewhat depreſſed and debilitated ; my propenſity to ſleep however, came on at the uſual hour, and as uſual was indulged in, my repoſe was found and unbroken.

Between May and July, I habitually breathed the gas, occaſionally three or four times a day for a week together ; at other periods, four or five times a week only.

The doſes were generally from ſix to nine quarts ; their effects appeared undiminiſhed by habit, and were hardly ever exactly ſimilar. Sometimes I had the feelings of intenſe intoxication, attended with but little pleaſure ; at other times, ſublime emotions connected with highly vivid ideas ; my pulſe was generally increaſed in fulneſs, but rarely in velocity.

The general effects of its operation upon my health and state of mind, are extremely difficult of defcription ; nor can I well difcriminate between its agency and that of other phyfical and moral caufes.

I flept much lefs than ufual, and previous to fleep, my mind was long occupied by vifible imagery. I had a conftant defire of action, a reftleffnefs, and an uneafy feeling about the præcordia analogous to the ficknefs of hope.

But perhaps thefe phænomena in fome meafure depended on the intereft and labour connected with the experimental inveftigation relating to the production of nitrous oxide, by which I was at this time inceffantly occupied.

My appetite was as ufual, and my pulfe not materially altered. Sometimes for an hour after the infpiration of the gas, I experienced a fpecies of mental indolence* pleafing rather than

---

* Mild phyfical pleafure is perhaps always deftructive to action. Almoft all our powerful voluntary actions, arife either from hope, fear, or defire; and the moft powerful from defire, which is an emotion produced by the coalefcence of hope or ideal pleafure with phyfical pain.

otherwife, and never ending in fiftlefnefs.

During the laft week in which I breathed it uniformly, I imagined that I had increafed fenfibility of touch : my fingers were pained by any thing rough, and the tooth edge produced from flighter caufes than ufual. I was certainly more irritable, and felt more acutely from trifling circumftances. My bodily ftrength was rather diminifhed than increafed.

At the latter end of July, I left off my habitual courfe of refpiration ; but I continued occafionally to breathe the gas, either for the fake of enjoyment, or with a view of afcertaining its operation under particular circumftances. In one inftance, when I had head-ache from indigeftion, it was immediately removed by the effects of a large dofe of gas ; though it afterwards returned, but with much lefs violence. In a fecond inftance, a flighter degree of head-ache was wholly removed by two dofes of gas.

The power of the immediate operation of the gas in removing intenfe phyfical pain, I had a very good opportunity of afcertaining,

number of experiments, that I took into my
lungs at every natural infpiration, about 13 cubic inches of air, and that I threw out of my
lungs at every expiration,* rather lefs than this
quantity; about 12¾ cubic inches.

The mean compofition of the 13 cubic inches
of air infpired, was

cub. in.

Nitrogene .. 9,5
Oxygene 3,4
Carbonic acid 0,1

That of the 12,7 of air expired

Nitrogene .. 9,3
Oxygene .. 2,2
Carbonic acid 1,2

Thefe refults I gained from more than 20 experiments, fo that I could not poffibly entertain
any doubt of this accuracy.

I found, by making a perfon obferve my refpirations when I was inattentive to the procefs,

---

* The diminution of air by fingle infpirations, was
particularly noticed by Dr. Goodwyn.

D d

that I made about 26 or 27 natural infpirations in a minute. So that calculating from the above eftimations, it would follow, that 31,6 cubic inches of oxygene were confumed, and 5,2 inches of nitrogene loft in refpiration every minute, whilft 26,6 cubic inches of carbonic acid were produced.

To collect the products of a great number of natural expirations fo as to afcertain whether their compofition correfponded with the above accounts, I proceeded in the following manner.

I faftened my lips tight on the mouth-piece of the exhaufted airholder, and fuffering my noftrils to remain open, infpired naturally through them, throwing the expired air through my mouth into the airholder.

In many experiments, I found that in about a half a minute, I made in this way 14 or 15 expirations. The mean quantity of air collected was 171 cubic inches, and confifted of

	cub. in.
Nitrogene ..	128
Oxygene ..	29
Carbonic acid ..	14

Comparing thefe refults with the former ones, we find the mean quantities of air refpired in equal terms rather lefs ; but the proportions of carbonic acid, nitrogene and oxygene in the refpired air, nearly identical.

e. To afcertain the changes effected in a given quantity of atmofpheric air by continued refpirations, I breathed after a compleat expiration, at temperature 63°, 161 cubic inches of air for near a minute, making in this time, 19 deep infpirations. After the compleat expiration, which was very carefully made, the gas filled a fpace nearly equal to 152 cubic inches, fo that 9 cubic inches of gas had difappeared.

The 152 cubic inches analifed, were found to confift of

	cub. in.
Nitrogene ..	111,6
Oxygene ..	23,
Carbonic acid,	17,4

The 161 cubic inches before infpiration, were compofed of

cub. in.

Nitrogene .. 117.0
Oxygene .. 42,4
Carbonic acid 1,6

But the refidual gas in the lungs before the experiment, was of different compofition from that remaining in the lungs after the experiment. Making corrections on account of this circumftance, as in fection IV. it appears that about 5,1 of nitrogene were abforbed in refpiration, 23,9 of oxygene confumed, and 12 of carbonic acid produced.

I repeated this experiment three times; in each experiment the diminution after refpiration, was nearly the fame; and the refidual gas making the neceffary allowances,. of fimilar compofition. So that fuppofing the exiftence of no fource of error in the experiments from which the quantity and compofition of the refidual gas of the lungs were eftimated in fection IV. the abforption of nitrogene by the venous blood, appears almoft demonftrated.

f. To compare the changes effected in atmospheric air by respiration of the smaller quadrupeds, with those in the experiments just detailed, I introduced into a jar of the capacity of 20 cubic inches filled with mercury in the mercurial trough, 15 cubic inches of atmospheric air which had been deprived of its carbonic acid by long exposure, to solution of potash.

Temperature being 64°, a healthy small mouse was quickly passed under the mercury into the jar, and suffered to rest on a very thin bit of cheese, which was admitted immediately after.

He continued for near 40 minutes without apparently suffering, occasionally raising himself on his hind legs. At the end of 50 minutes, he was lying on his side, and in 55 minutes was apparently dying. He was now carefully taken out through the mercury by the tail, and exposed before the fire, where he soon recovered. After the cheese had been carefully removed, the gas in the jar filled a space nearly equal to 14 cubic inches; so that a diminution of a cubic inch had taken place. Thefe 14 cubic inches analifed, were found to confift of

cub. in.

Carbonic acid .,	2,0
Oxygene	1,4
Nitrogene ...	10,6

The 15 cubic inches before the experiment, confifted of

cub. in.

Oxygene ..	4
Nitrogene ..	11

Hence it appeared, that 2,6 cubic inches of oxygene had been confumed, 2 cubic inches of carbonic acid produced, and about 0,4 of nitrogene loft.

The relation between the quantities of oxygene confumed in this experiment, and the carbonic acid produced, are nearly the fame as that of thofe in the experiments juft detailed; but the quantity of nitrogene loft is much fmaller.

VII. *Respiration of Oxygene.*

. The gases before and after respiration, were analised in these experiments in the manner described in the last section, except that 3 : of nitrous gas were always employed to one of oxygene.

E. I. At temperature 53°, after a full forced respiration, I respired in the mercurial airholder, for half a minute, 102 cubic inches of oxygene, making seven very long and deep inspirations. After the compleat expiration, the gases filled a space equal to 93 cubic inches; these 93 cubic inches analised, were found to consist of

	cub. in.
Carbonic acid ..	5,9
Nitrogene , ... :	33,8
Oxygene	53,3

The 102 cubic inches before the experiment, were composed of

	cub. in.
Oxygene ..	78
Nitrogene. ..	24

The refidual gas in the lungs before the experiment, was 32 cubic inches, and compofed of
about 23 nitrogene, 4,1 carbonic acid, and 4,9
oxygene, Section IV. The refidual gas after
expiration, was compofed of 18,2 oxygene,
2 carbonic acid, and 11,8 nitrogene.

Hence the whole of the gas in the lungs and
airholder before infpiration, was 134 cubic
inches, compofed of

	cub. in.
Oxygene ...	82,9
Nitrogene ...	47,0
Carbonic acid ...	4,1

And after refpiration, 125 cubic inches, confifting of

	cub. in.
Oxygene	71,5
Nitrogene	45,6
Carbonic acid ..	7,9

So that comparing the quantities, it appears,
that 11,4 of oxygene and 1,4 of nitrogene,
were confumed in this experiment, and 3,8 of
carbonic acid produced.

I was much furprifed at the fmall quantity of oxygene that had been confuméd-in this experiment. This quantity was lefs than, that expended during the refpiration of atmofpheric air for half a minute : the portion of carbonic acid revolved was likewife fmaller. I could detect no fource of inaccuracy, and it was difficult to fuppofe that the greater depth and fulnefs of the infpirations could make any difference.

E. 2. I now refpired at the fame temperature, after a full expiration, 162 cubic inches of gas, compofed of 133 oxygene and 29 nitrogene for two minutes, imitating as much as poffible, the natural refpiration. After the experiment, they filled a fpace equal to 123 cubic inches. And when the analyfis and calculations had been made as in the laft experiment, it appeared that 57 cubic inches of oxygene, and 2 of nitrogene had been abforbed; whilft 21 cubic inches of carbonic acid had been formed.

Now from the eftimations in the laft fection,

it appears that 63 cubic inches of oxygene are confumed, and about 52 cubic inches of carbonic acid produced every two minutes during the natural refpiration of common air. So that fuppofing the experiment accurate, 6 cubic inches of oxygene lefs are abforbed, and 30 cubic inches lefs of carbonic acid produced every minute, when oxygene nearly pure is refpired; than when atmofpheric air is refpired.

Both thefe experiments were made in the morning, at a time when I was in perfect health ; fo that there could be apparently no fource of error from accidental circumftances.

The uncommon and unexpected nature of the refults, made me however, very fceptical; concerning them ; and before I would draw any inferences, I refolved to afcertain the comparative confumption of atmofpheric air and oxygene by the fmaller quadrupeds, for which purpofe, I made the following experiment.

E. 3. Of two ftrong and healthy fmall mice, apparently of the fame breed, and exactly fimilar.

One was introduced into a jar containing 10 cubic inches and half of oxygene, and 3 cubic inches of nitrogene, and made to rest on a bit of cheese.

The other was introduced into a jar containing fifteen cubic inches and half of atmospheric air, and made to rest in the same manner on cheese.

The mouse in oxygene began apparently to suffer in about half an hour, and occasionally panted very much; in about an hour he lay down on his side as if dying. The jars were often agitated, that the gases might be well mingled.

The mouse in atmospheric air became very feeble in 40 minutes, and at the end of 50 minutes was taken out through the mercury alive, but unable to stand.

The mouse in oxygene was taken out in the same manner after an hour and quarter, alive, but motionless, and breathing very deeply.

The gas in the jars was examined. That in the oxygene jar filled a space exactly equal to

12,7 cubic inches, and analifed, was found to confift of 1,7 carbonic acid, 2,6 of nitrogene, and 8,4 of oxygene. So that abfolutely, 2,1 cubic/inches of oxygene and ,4 of nitrogene had been confumed, and 1,7 of carbonic acid produced.

The gas in the atmofpheric air jar was diminifhed nearly to 14,4, and confifted of 2,1 carbonic acid, 1,4 oxygene; and 10,9 nitrogene. So that 2,7 of oxygene and ,5 of nitrogene, had been confumed by the moufe; and 2,1 of carbonic acid produced.

Hence it appears, that the moufe in atmofpheric air confumed nearly one-third more oxygene and produced nearly one-fourth more carbonic acid in refpiration in 55 minutes, than the other in an hour and quarter in oxygene. And if we confider the perpetual diminution of the oxygene of the atmofpheric air; from which at laft it became almoft incapable of fupporting the life of the animal; we may conclude, that the quantity of oxygene confumed by it, had the air been perpetually renovated, would have been much more confiderable.

I defign very fhortly, to repeat thefe experiments, and to make others on the comparative confumption of oxygene and atmofpheric air, by the larger quadrupeds. Whatever may be the refults, I hope to be able to afcertain from them, why pure oxygene is incapable of fupporting life.

VIII. *Obfervations on the changes effected in the blood, by atmofpheric air and oxygene.*

From the experiments of Mr. Cigna and Dr. Prieftley,* it appears that the coagulum of the venous blood becomes florid at its furface when expofed to the atmofphere, though covered and defended from the immediate contact of air by a very thick ftratum of ferum.

* Dr. Prieftley found that it likewife became florid at the furface when covered by milk ; but that it underwent little or no alteration of color under water and moft other fluids.—Vol. 3. p. 372.

᾿ Hence it is evident, that ferum is capable of diffolving either the whole compound atmofphe- ric air, or the oxygene of it.

Suppofing what indeed is moft probable from numerous analogies, that it diffolves the whole compound ; it would follow, that the coloring of the coagulum of blood under ferum, depended upon the decompofition of the atmofpheric air condenfed in the ferum, the oxygene† of it, combining with the red particles, and the nitro- gene either remaining diffolved in the fluid, or being liberated through it into the atmofphere.

Now the circulating blood confifts of red par- ticles, floating in and diffufed through ferum and coagulable lymph.

† There are many analogous decompofitions. Dr. Prieftley noticed (and I have often made the obfervation) that green oxide of iron, or the precipitate from pale green fulphate of iron by cauftic alkali, became red at the furface, when covered by a thick ftratum of water. In my experiments on the green muriate and fulphate of iron, I obferved that part of fome dark oxide of iron which was at the bottom of a trough of water 9 inches deep, became red at the furface nearly in the fame time as another portion of the fame preci- pitation that was expofed to the atmofphere. This oxyge- nation muft depend upon the decompofition of atmofpheric air conftantly diffolved by the water.

In natural refpiration, the red particles are rendered of a brighter tinge during the paffage of the blood through the pulmonary veins. And as we have feen in the laft fections, during refpiration atmofpheric air is decompofed; all the oxygene of it confumed, *apparently* a fmall portion of the nitrogene loft, and a confiderable quantity of carbonic acid produced.

It feems therefore reafonable to fuppofe, that the whole compound atmofpheric air paffing through the moift coats of the veffels is firft diffolved by the ferum of the venous blood, and in its condenfed ftate, decompofed by the affinity of the red particles for its oxygene; the greater part of the nitrogene being liberated unaltered; but a minute portion of it poffibly remaining condenfed in the ferum and coagulable lymph, and paffing with them into the left chamber of the heart.

From the experiments on the refpiration of nitrous oxide and hydrogene, it appears that a certain portion of the carbonic acid produced in refpiration, is evolved from the venous blood;

but as a much greater quantity is generated during the refpiration of common air and oxygene, than during that of hydrogene in equal times, it is not impoffible but that fome portion of it may be formed by the combination of charcoal in the red particles with the oxygene diffolved in the ferum ; but this can only be determined by farther experiments.

Suppofing that no part of the water evolved in folution by the expired gas of common air is formed immediately in refpiration, it will follow that a very confiderable quantity of oxygene muft be conftantly *combined* with the red particles, even allowing the confumption of a certain portion of it to form carbonic acid ; for the carbonic acid evolved, rarely amounts to more than three-fourths of the volume of the oxygene confumed.

Perhaps the ferum of the blood is capable of diffolving a larger quantity of atmofpheric air than of pure oxygene. On this fuppofition, it would be eafy to explain the fmaller confumption of oxygene in the experiments in the laft fection.

In cutting one of the unlucky teeth called dentes fapientiæ, I experienced an extenfive inflammation of the gum, accompanied with great pain, which equally deftroyed the power of repofe, and of confiftent action.

On the day when the inflammation was moft troublefome, I breathed three large dofes of nitrous oxide. The pain always diminifhed after the firft four or five infpirations ; the thrilling came on as ufual, and uneafinefs was for a few minutes, fwallowed up in pleafure. As the former ftate of mind however returned, the ftate of organ returned with it ; and I once imagined that the pain was more fevere after the experiment than before.

In Auguft, I made many experiments with a view of afcertaining whether any analogy exifted between the fenfible effects of the different gafes which are fooner or later fatal to life when refpired, and thofe of nitrous oxide.

I refpired four quarts of Hydrogene* nearly

* Pure hydrogene has been often refpired by different Philofophers, particularly by Scheele, Fontana, and the adventurous and unfortunate Rofier.

pure produced from zinc and muriatic acid, for
near a minute; my lungs being previoufly exhaufted and my noftrils carefully clofed. The
firft fix or feven infpirations produced no fenfations whatever; in half a minute, I perceived
a difagreeable oppreffion of the cheft, which
obliged me to refpire very quickly ; this oppreffion gradually increafed, till at laft the pain
of fuffocation compelled me to leave off breathing. I felt no giddinefs during or after the
experiment ; my pulfe was rendered feebler and
quicker; and a by-ftander informed me that
towards the laft, my cheeks became purple.

In a fecond experiment, when the hydrogene was procured from iron and diluted fulphuric acid, I was unable to refpire it for fo
long as three quarters of a minute ; a tranfient
giddinefs and mufcular debility were produced;
the pulfe was rendered very feeble, and the pain
of fuffocation was greater than before.

I breathed three quarts of Nitrogene mingled
with a very fmall portion of carbonic acid, for
near a minute. It produced no alteration in

my fenfations for the firft twenty feconds ; then the painful fenfe of fuffocation gradually came on, and increafed rapidly in the laft quarter of the minute, fo as to oblige me to defift from the experiment. My pulfe was rendered feebler and quicker. I felt no affection whatever in the head.

Mr. Watt's obfervations on the refpiration of diluted Hydrocarbonate by men, and Dr. Beddoes's experiments on the deftruction of animals by pure hydrocarbonate, proved that its effects were highly deleterious.

As it deftroyed life apparently by rendering the mufcular fibre inirritable without producing any previous excitement, I was anxious to compare its fenfible effects with thofe of nitrous oxide, which at this time I believed to deftroy life by producing the higheft poffible excitement, ending in læfion of organifation.

In the firft experiment, I breathed for near a minute, three quarts of hydrocarbonate mingled with nearly two quarts of atmofpheric air.*

---

* I believe it had never been breathed before by any individual, in a ftate fo little diluted.

It produced a flight giddinefs and pain in the head, and a momentary lofs of voluntary power : my pulfe was rendered much quicker and feebler. Thefe effects however, went off in five minutes, and I had no return of giddinefs.

· Emboldened by this trial, in which the feelings were not unlike thofe I experienced in the firft experiments on nitrous oxide, I refolved to breathe pure hydrocarbonate.

For this purpofe, I introduced into a filk bag, four quarts of gas nearly pure, which was carefully produced from the decompofition of water by charcoal an hour before, and which had a very ftrong and difagreeable fmell.

My friend, Mr. James Tobin, Junr. being prefent, after a forced exhauftion of my lungs, the nofe being accurately clofed, I made three infpirations and expirations of the hydrocarbonate. The firft infpiration produced a fort of numbnefs and lofs of feeling in the cheft and about the pectoral mufcles. After the fecond infpiration, I loft all power of perceiving external things, and had no diftinct fenfation except a terrible oppreffion on the cheft. During the third expiration, this feeling difappeared, I feemed finking into annihilation, and had juft power enough to drop the mouth-piece from my unclofed lips. A fhort interval muft have paffed during which I refpired common air, before the objects about me were diftinguifhable. On recollecting myfelf, I faintly articulated, " *I do not think I fhall die.*" Putting my finger on the wrift, I found my pulfe thread-like and beating with exceffive quicknefs.

In lefs than a minute, I was able to walk, and the painful oppreffion on the cheft directed me to the open air.

After making a few fteps which carried me to the garden, my head became giddy, my knees trembled, and I had juft fufficient voluntary power to throw myfelf on the grafs. Here the painful feeling of the cheft increafed with fuch violence as to threaten fuffocation. At this moment, I afked for fome nitrous oxide. Mr. Dwyer brought me a mixture of oxygene and nitrous oxide. I breathed this for a minute, and *believed* myfelf relieved. In five minutes, the painful feelings began gradually to diminifh. In an hour they had nearly difappeared, and I felt only exceffive weaknefs and a flight fwimming of the head. My voice was very feeble and indiftinct. This was at two o'clock in the afternoon.

I afterwards walked flowly for about half an hour, with Mr. Tobin, Junr. and on my return, was fo much ftronger and better, as to believe that the effects of the gas had difappeared; though my pulfe was 120, and very feeble. I continued without pain for near three quarters of an hour; when the giddinefs returned with fuch violence as to oblige me to lie on the bed; it was accompanied with naufea, lofs of memory, and deficient fenfation. In about an hour and half, the giddinefs went off, and was fucceeded by an excruciating pain in the forehead and between the eyes, with tranfient pains in the cheft and extremities.

Towards night thefe affections gradually diminifhed., At ten,† no difagreeable feeling except weaknefs remained. I flept found, and awoke in the morning very feeble and very hungry. No recurrence of the fymptoms took place, and I had nearly recovered my ftrength by the evening.

I have been minute in the account of this experiment becaufe it proves, that hydrocarbonate acts as a fedative, i. e. that it produces diminution of vital action, and debility, without previoufly exciting. There is every reafon to believe, that if I had taken four or five infpirations inftead of three, they would have deftroyed life immediately without producing any painful fenfation. Perhaps moft of the uneafy feelings after the experiment, were connected with the return of the healthy condition of organs.*

† I ought to obferve, that between eight and ten, I took by the advice of Dr. Beddoes, two or three dofes of diluted nitric acid.

* By whatever caufe the exhauftion of organs is produced, pain is almoft uniformly connected with their returning health. Pain is rarely ever perceived in limbs debilitated

About a week after this experiment, I attempted to refpire Carbonic acid,. not being at the time acquainted with the experiments of Rofier.

I introduced into a filk bag four quarts of well wafhed carbonic acid produced from carbonate of ammoniac* by heat, and after a compleat voluntary exhauftion of my lungs, attempted to infpire it. It tafted ftrongly acid in the mouth and fauces, and produced a fenfe of burning at the top of the uvula. In vain I made powerful voluntary efforts to draw it into the windpipe ; at the moment that the epiglottis was raifed a little, a painful ftimulation was induced, fo as to clofe it fpafmodically on the glottis ; and thus in repeated trials I was prevented from taking a fingle particle of carbonic acid into my lungs.

by fatigue till after they have been for fome hours at reft. Pain is uniformly connected with the recovery from the debility induced by typhus, often with the recovery from that produced by the ftimulation of opium and alcohol.

* Carbonic acid is produced in this way in a high ftate of purity, and with great readinefs.

· I tried to breathe a mixture of two quarts of common air and three of carbonic acid, without fuccefs; it ftimulated the epiglottis nearly in the fame manner as pure carbonic acid, and was perfectly non-refpirable.

I found that a mixture of three quarts of carbonic acid with feven of common air was refpirable, I breathed it for near a minute. At the time, it produced a flight degree of giddinefs, and an inclination to fleep. Thefe effects however, very rapidly difappeared after I had ceafed to breathe,* and no other affections followed.

During the courfe of experiments on nitrous oxide, I feveral times breathed Oxygene procured from manganefe by heat, for from three to five minutes.

In refpiring eight or ten quarts; for the firft

---

* Carbonic acid poffeffes no action on arterial blood. Hence perhaps, its flight effects when breathed mingled with large quantities of common air. Its effects are very marked upon venous blood! If it were thrown forcibly into the lungs of animals, the momentary application of it to the pulmonary venous blood would probably deftroy life.

two or three minutes I could perceive no effects.
Towards the end, even when I breathed very
flowly, my respiration became oppressed, and
I felt a fenfation analogous to that produced by
the want of frefh air; though but little of the
oxygene had been confumed.

In one experiment when I breathed from and
into a bag containing 20 quarts of oxygene for
near fix minutes; Dr. Kinglake felt my pulfe,
and found it not altered in velocity, but rather
harder than before. I perceived no effects but
thofe of oppreffion on the cheft*.

* In a conversation with Mr. Watt, relating to the powers of gafes, that excellent philofopher told me he had for
fome time entertained a fufpicion, that the effects attribu-
ted to oxygene produced from manganefe by heat, in fome
meafure depended upon nitrous acid fufpended in the gas,
formed during ignition by the union of fome of the oxygene
of the manganefe with nitrogene likewife condenfed in it.
In the courfe of experiments on nitrous acid, detailed in
Refearch I. made in September, October, and December,
1799, I feveral times experienced a fevere oppreffion on the
cheft and difficulty of refpiration, not unanalogous to that
produced by oxygene, but much more violent, from
breathing an atmofphere loaded with nitrous acid vapour.
This fact feemed to confirm Mr. Watt's fufpicion. I con-

Having obferved in my experiments upon venous blood, that Nitrous gas rendered that fluid of a purple tinge, very like the color generated in it by nitrous oxide; and finding no painful effects produced by the application of nitrous gas to the bare muscular fibre; I began to imagine that this gas might be breathed with impunity, provided it were poffible in any way to free the lungs of common air before infpiration, fo as to prevent the formation of nitrous acid.

On this fuppofition, during a fit of enthufiafm produced by the refpiration of nitrous oxide, I refolved to endeavour to breathe Nitrous gas.

114 cubic inches of nitrous gas were introduced into the large mercurial airholder; two

fefs, however, that I have never been able to detect any fmell of nitrous acid, either by means of my own organs or thofe of others, during the production of oxygene; when the gas is fuffered to pafs into the atmofphere. The oxygene breathed in the experiments detailed in the text, had been for fome days in contact with water.

fmall filk bags of the capacity of feven quarts were filled with nitrous oxide.

After a forced exhauftion of my lungs, my nofe being accurately clofed, I made three in-fpirations and expirations of nitrous oxide in one of the bags, to free my lungs as much as poffible from atmofpheric oxygene.; then, after a full expiration of the nitrous oxide, I transferred my mouth from the mouth-piece of the bag to that of the airholder, and turning the ftop-cock, attempted to infpire the nitrous gas.—In paffing through my mouth and fauces, it tafted aftringent and highly difagreeable; it occafioned a fenfe of burning in the throat, and produced a fpafm of the epiglottis fo painful as to oblige me to defift inftantly from attempts to infpire it. After moving my lips from the mouth-piece, when I opened them to infpire common air, aëriform nitrous acid was inftantly formed in my mouth, which burnt the tongue and palate, injured the teeth, and produced an inflammation of the mucous membrane which lafted for fome hours.

As after the refpiration of nitrous oxide in
the experiments in the laft Refearch, a fmall
portion of the refidual atmofpheric air remained
in the lungs, mingled with the gas, after forced
expiration ; it is moft probable that a minute
portion of nitrous acid was formed in this experiment, when the nitrous gas was taken into
the mouth and fauces; which might produce
its ftimulating properties. If fo, perhaps I
owe my life to the circumftance ; for fuppofing I had taken an infpiration of nitrous
gas, and even that it had produced no
pofitive effects, it is highly improbable, that
by breathing nitrous oxide, I fhould have freed
my lungs from it, fo as to have prevented the
formation of nitrous acid when I again infpired
common air. I never defign again to attempt
fo rafh an experiment.

In the beginning of September I often refpired nitrous oxide mingled with different proportions of common air or oxygene. The
effects produced by the diluted gas were much
lefs violent than thofe produced by pure nitrous

oxide. They were generally pleasant : the thrilling was not often perceived, but a sense of exhilaration was almost constant.

Between September and the end of October, I made but few experiments on respiration, almost the whole of my time being devoted to chemical experiments on the production and analysis of nitrous oxide.

At this period my health being somewhat injured by the constant labour of experimenting, and the perpetual inhalation of the acid vapours of the laboratory, I went into Cornwal ; where new associations of ideas and feelings, common exercise, a pure atmosphere, luxurious diet and moderate indulgence in wine, in a month restored me to health and vigor.

Nov. 27th. Immediately after my return, being fatigued by a long journey, I respired nine quarts of nitrous oxide, having been precisely thirty-three days without breathing any. The feelings were different from those I had experienced in former experiments. After the first six or seven inspirations, I gradually began to lofe the perception of external things, and a vivid and intenfe recollection of fome former experiments paffed through my mind, fo that I called out " *what an amazing concatenation of ideas !*" I had no pleafurable feeling whatever, I ufed no mufcular motion, nor did I feel any difpofition to it ; after a minute, when I made the note of the experiment, all the uncommon fenfations had vanifhed ; they were fucceeded by a flight forenefs in one of the arms and in the leg : in three minutes thefe affections likewife difappeared.

From this experiment I was inclined to fuppofe that my newly acquired health had diminifhed my fufceptibility to the effects of the gas. About ten days after, however, I had an opportunity of proving the fallacy of this fuppofition.

Immediately after a journey of 126 miles, in which I had no fleep the preceding night, being much exhaufted, I refpired feven quarts of gas for near three minutes. It produced the ufual pleafurable effects, and flight mufcular motion.

I continued exhilarated for fome minutes afterwards: but in half an hour found myfelf neither
more or lefs exhaufted than before the experiment. I had a great propenfity to fleep.

I repeated the experiment four or five times
in the following week, with fimilar effects. My
fufceptibility was certainly not diminifhed. I
even thought that I was more affected than formerly by equal dofes.

Though, except in one inftance, when indeed
the gas was impure, I had experienced no decifive
exhauftion after the excitement from nitrous oxide, yet ftill I was far from being fatisfied that it
was unanalogous to ftimulants in general.—
No experiment had been made in which the
excitement from nitrous oxide had been kept up
for fo great a length of time and carried to fo
great an extent as that in which it is uniformly
fucceeded by exceffive debility under the agency
of other powers.

It occurred to me, that fuppofing nitrous oxide to be a ftimulant of the common clafs, it
would follow that the debility produced in con-

fequence of exceffive ftimulation by a known
agent, ought to be *increafed* after excitement
from nitrous oxide.*

To afcertain whether this was the cafe, I
made on December 23d, at four P. M. the
following experiment. I drank a bottle of
wine in large draughts in lefs than eight minutes. Whilft I was drinking, I perceived a
fenfe of fulnefs in the head, and throbbing of
the arteries, not unanalogous to that produced in
the firft ftage of nitrous oxide excitement.
After I had finifhed the bottle, this fulnefs increafed, the objects around me became dazzling,
the power of diftinct articulation was loft, and
I was unable to walk fteadily. At this moment
the fenfations were rather pleafurable than otherwife, the fenfe of fulnefs in the head foon however increafed fo as to become painful, and in

---

* In the fame manner as the debility from intoxication
by two bottles of wine is increafed by a third.

lefs than an hour I funk into a ftate of infenfi-
bility.*

In this fituation I muft have remained for
two hours or two hours and half.

I was awakened by head-ache and painful
naufea. The naufea continued even after the
contents of the ftomach had been ejected. The
pain in the head every minute increafed; I was
neither feverifh or thirfty; my bodily and mental debility were exceffive, and the pulfe feeble
and quick.

In this ftate I breathed for near a minute and
half five quarts of gas, which was brought to
me by the operator for nitrous oxide; but as it
produced no fenfations whatever, and apparently
rather increafed my debilty, I am almoft convinced that it was from fome accident, either
common air, or very impure nitrous oxide.

Immediately after this trial, I refpired 12 quarts

---

* I ought to obferve that my ufual drink is water, that
I had been little accuftomed to take wine or fpirits, and
had never been compleatly intoxicated but once before in
the courfe of my life. This will account for the powerful
effects of a fingle bottle of wine.

of oxygene for near four minutes. It produced no alteration in my fenfations at the time ; but immediately after I imagined that I was a little exhilarated.

The head-ache and debility ftill however continuing with violence, I examined fome nitrous oxide which had been prepared in the morning, and finding it very pure, refpired feven quarts of it for two minutes and half.

I was unconfcious of head-ache after the third infpiration ; the ufual pleafurable thrilling was produced, voluntary power was deftroyed, and vivid ideas rapidly paffed through my mind ; I made ftrides acrofs the room, and continued for fome minutes much exhilarated. Immediately after the exhilaration had difappeared, I felt a flight return of the head-ache ; it was connected with tranfient naufea. After two minutes, when a fmall quantity of acidified wine had been thrown from the ftomach, both the naufea and head-ache difappeared ; but languor and depreffion not very different in degree from thofe exifting before the experiment, fucceeded.

They however, gradually went off before bed time. I slept found the whole of the night except for a few minutes, during which I was kept awake by a trifling head-ache. In the morning, I had no longer any debility. No head-ache or giddiness came on after I had arisen, and my appetite was very great.

This experiment proved, that debility from intoxication was not increased by excitement from nitrous oxide. The head-ache and depression, it is probable, would have continued longer if it had not been administered. Is it not likely that the slight naufea following the effects of the gas was produced by new excitability given to the stomach?

To ascertain with certainty, whether the most extensive action of nitrous oxide compatible with life, was capable of producing debility, I resolved to breathe the gas for such a time and in such quantities, as to produce excitement equal in duration and superior in intensity to that occasioned by high intoxication from opium or alcohol.

To habituate myfelf to the excitement, and to carry it on gradually

On December 26th, I was inclofed in an air-tight breathing-box;* of the capacity, of about 9 cubic feet and half, in the prefence of Dr. Kinglake.

After I had taken a fituation in which I could by means of a curved thermometer inferted under the arm, and a ftop-watch, afcertain the alterations in my pulfe and animal heat, 20 quarts of nitrous oxide were thrown into the box.

For three minutes I experienced no alteration in my fenfations, though immediately after the introduction of the nitrous oxide the fmell and tafte of it were very evident.†

In four minutes I began to feel a flight glow.

---

* The plan of this box was communicated by Mr. Watt. An account of it will be detailed in the *Refearches*.

† The nitrous oxide was too diluted to act much; it was mingled with near 32 times its bulk of atmofpheric air.

in the cheeks, and a generally diffused warmth over the cheft, though the temperature of the box was not quite 50°. I had neglected to feel my pulfe before I went in; at this time it was 104 and hard, the animal heat was 98°. In ten minutes the animal heat was near 99°, in a quarter of an hour 99.5°, when the pulfe was 102, and fuller than before.

At this period 20 quarts more of nitrous oxide were thrown into the box, and well-mingled with the mafs of air by agitation.

In 25 minutes the animal heat was 100°, pulfe 124. In 30 minutes, 20 quarts more of gas were introduced.

My fenfations were now pleafant; I had a generally diffufed warmth without the flighteft moifture of the fkin, a fenfe of exhilaration fimilar to that produced by a fmall dofe of wine, and a difpofition to mufcular motion and to merriment.

In three quarters of an hour the pulfe was 104, and animal heat not 99,5°, the tempera-ture of the chamber was 64°. The pleafurable feelings continued to increafe, the pulfe became fuller and flower, till in about an hour it was 88°, when the animal heat was 99°.

20 quarts more of air were admitted. I had now a great difpofition to laugh, luminous points feemed frequently to pafs before my eyes, my hearing was certainly more acute and I felt a pleafant lightnefs and power of exertion in my mufcles. In a fhort time the fymptoms became ftationary; breathing was rather oppreffed, and on account of the great defire of action, reft was painful.

I now came out of the box, having been in precifely an hour and quarter.

The moment after, I began to refpire 20 quarts, of unmingled nitrous oxide. A thrilling ex- tending from the cheft to the extremities was almoft immediately produced. I felt a fenfe of tangible extenfion highly pleafurable in every limb; my vifible impreffions were dazzling and apparently magnified, I heard diftinctly every found in the room and was perfectly aware

of my situation.* By degrees as the pleasurable
sensations increased, I lost all connection with
external things; trains of vivid visible images
rapidly passed through my mind and were connected with words in such a manner, as to produce perceptions perfectly novel. I existed in
a world of newly connected and newly modified
ideas. I theorised; I imagined that I made
discoveries. When I was awakened from this
semi-delirious trance by Dr. Kinglake, who
took the bag from my mouth, Indignation
and pride were the first feelings produced by
the sight of the persons about me. My emotions
were enthusiastic and sublime; and for a minute
I walked round the room perfectly regardless
of what was said to me. As I recovered my
former state of mind, I felt an inclination to
communicate the discoveries I had made during
the experiment. I endeavoured to recall the
ideas, they were feeble and indistinct; one
collection of terms, however, presented itself:

* In all these experiments after the first minute, my
cheeks became purple.

and with the moſt intenſe belief; and prophetic manner, I exclaimed to Dr. Kinglake, *" Nothing exiſts but thoughts!—the univerſe is compoſed of impreſſions, ideas, pleaſures and pains !"*

About three minutes and half only, had elap-ſed during this experiment, though the time as meaſured by the relative vividneſs of the recol-lected ideas, appeared to me much longer.

Not more than half of the nitrous oxide was conſumed. After a minute, before the thrilling of the extremities had diſappeared, I breathed the remainder. Similar ſenſations were again produced ; I was quickly thrown into the plea-ſurable trance, and continued in it longer than before. For many minutes after the experiment, I experienced the thrilling in the extremities, the exhilaration continued nearly two hours. For a much longer time I experienced the mild enjoyment before deſcribed, connected with indolence ; no depreſſion or feeblenefs followed. I ate my dinner with great appetite and found myſelf lively and diſpoſed to action immediately after. I paſſed the evening in executing experiments. At night I found myself unusually cheerful and active; and the hours between eleven and two, were spent in copying the foregoing detail from the common-place book and in arranging the experiments. In bed I enjoyed profound repose. When I awoke in the morning, it was with confcioufness of pleafurable exiftence, and this confcioufnefs more or lefs, continued through the day.

Since December, I have very often breathed nitrous oxide. My fufceptibility to its power is rather increafed than diminifhed. I find fix quarts a full dofe, and I am rarely able to refpire it in any quantity for more than two minutes and half.

The mode of its operation is fomewhat altered. It is indeed very different at different times.

I am fcarcely ever excited into violent mufcular action, the emotions are generally much lefs intenfe and fublime than in the former experiments, and not often connected with thrilling in the extremities.

When troubled with indigeftion, I have been two or three times unpleafantly affected after the excitement of the gas. Cardialgia, eructations and unpleafant fulnefs of the head were produced.

I have often felt very great pleafure when breathing it alone, in darknefs and filence, occupied only by ideal exiftence. In two or three inftances when I have breathed it amidft noife, the fenfe of hearing has been painfully affected even by moderate intenfity of found. The light of the fun has fometimes been difagreeably dazzling. I have once or twice felt an uneafy fenfe of tenfion in the cheeks and tranfient pains in the teeth.

Whenever I have breathed the gas after excitement from moral or phyfical caufes, the delight has been often intenfe and fublime.

On May 5th, at night, after walking for an hour amidft the fcenery of the Avon, at this period rendered exquifitely beautiful by bright moonfhine; my mind being in a ftate of agreeable feeling, I refpired fix quarts of newly prepared nitrous oxide.

The thrilling was very rapidly, produced. The objects around me were perfectly diftinct, and the light of the candle not as ufual dazzling. The pleafurable fenfation was at firft local and perceived in the lips and about the cheeks. It gradually however, diffufed itfelf over the whole body, and in the middle of the experiment was for a moment fo intenfe and pure as to abforb exiftence. At this moment, and not before; I loft confcioufnefs; it was however, quickly reftored, and I endeavoured to make a by-ftander acquaint- ed with the pleafure I experienced by laughing and ftamping. I had no vivid ideas. The thrilling and the pleafurable feeling continued for many minutes; I felt two hours afterwards, a flight recurrence of them, in the intermediate ftate between fleeping and waking; and I had du- ring the whole of the night, vivid and agreeable dreams. I awoke in the morning with the feeling of reftlefs energy, or that defire of action connected with no definite object, which I had often experienced in the course of experiments in 1799.

I have two or three times since respired nitrous oxide under similar circumstances; but never with equal pleasure.

During the last fortnight, I have breathed it very often; the effects have been powerful and the sensations uncommon; but pleasurable only in a slight degree.

I ought to have observed that a desire to breathe the gas is always awakened in me by the sight of a person breathing, or even by that of an air-bag or an air-holder.

I have this day, June 5th, respired four large doses of gas. The first two taken in the morning acted very powerfully; but produced no thrilling or other pleasurable feelings. The effects of the third breathed immediately after a hearty dinner were pleasant, but neither intense or intoxicating. The fourth was respired at night in darkness and silence after the occurrence of a circumstance which had produced some anxiety. This dose affected me powerfully and pleafantly.; a flight thrilling in the extremities was produced ; an exhilaration continued for fome time, and I have had but little return of uneafinefs. 11 P. M.

From the nature of the language of feeling, the preceding detail contains many imperfections; I have endeavoured to give as accurate an account as poffible of the ftrange effects of nitrous oxide, by making ufe of terms ftanding for the moft fimilar common feelings.

We are incapable of recollecting pleafures and pains of fenfe.* It is impoffible to reafon concerning them, except by means of terms which have been affociated with them at the moment of their exiftence, and which are afterwards called up amidft trains of concomitant ideas.

---

* Phyfical pleafure and pain generally occur connected with a compound impreffion, i. e. an organ and fome object. When the idea left by the compound impreffion, is called up by being linked accidentally to fome other idea or impreffion, no recurrence, or the flighteft poffible, of the pleafure or pain in any form will take place. But when the compound impreffion itfelf exifts *without* the phyfical pleafure or pain, it will awaken ideal or intellectual

When pleasures and pains are new or connected with new ideas, they can never be intelligibly detailed unless associated during their existence with terms standing for analogous feelings.

I have sometimes experienced from nitrous oxide, sensations similar to no others, and they have consequently been indescribable. This has been likewise often the case with other persons. Of two paralytic patients who were asked what they felt after breathing nitrous

pleasure or pain, i. e. hope or fear. So that physical pleasure and pain are to hope and fear, what impressions are to ideas. For instance, assuming no accidental association, the child does not fear the fire before he is burnt. When he puts his finger to the fire he feels the physical pain of burning, which is connected with a visible compound impression, the fire and his finger. Now when the compound idea of the fire and his finger, left by the compound impression are called up by his mother, saying, " *You have burnt your finger*," nothing like fear or the pain of burning is connected with it. But when the finger is brought near the fire, i. e. when the compound impression again exists, the ideal pain of burning or the passion of fear is awakened, and it becomes connected with those very actions which removed the finger from the fire.

oxide, the firſt anſwered, "*I do not know how, but very queer.*" The ſecond ſaid, "*I felt like the ſound of a harp.*" Probably in the one caſe, no analogous feelings had ever occurred. In the other, the pleaſurable thrillings were ſimilar to the ſenſations produced by muſic ; and hence, they were connected with terms formerly applied to muſic.

DIVISION II.

DETAILS *of the* EFFECTS *produced by the* RES-
PIRATION *of* NITROUS OXIDE *upon different*
INDIVIDUALS *furnished by* THEMSELVES.

THE experiments related in the following
details, were made in the Medical Pneumatic
Inftitution.

Abftracts from many of them have been
publifhed by Dr. Beddoes.*

I. *Detail of* Mr. J. W. TOBIN.

Having feen the remarkable effects produced
on Mr. Davy, by breathing nitrous oxide, the
18th of April; I became defirous of taking fome.

A day or two after I breathed 2 quarts of this

---

* Notice of fome Obfervations made at the Medical
Pneumatic Inftitution.

gas, returning it back again into the fame bag, after two or three infpirations, breathing became difficult, and I occafionally admitted common air into my lungs. While the refpiration was continued, my fenfations became more pleafant. On taking the bag from my mouth, I ftaggered a little, but felt no other effect.

On the fecond time of making the experiment, I took nearly four quarts, but ftill found it difficult to continue breathing long, though the air which was left in the bag was far from being impure.

The effects however, in this cafe, were more ftriking than in the former. Increafed mufcular action was accompanied by very pleafurable feelings, and a ftrong defire to continue the infpiration. On removing the bag from my mouth, I laughed, ftaggered, and attempted to fpeak, but ftammered exceedingly, and was utterly unable to pronounce fome words. My ufual flate of mind, however, foon returned.

On the 29th, I again breathed four quarts. The pleafant feelings produced at firft, urged

me to continue the infpiration with great eagernefs. Thefe feelings however, went off towards
the end of the experiment, and no other effects
followed. The gas had probably been breathed
too long, as it would not fupport flame. I then
propofed to Mr. Davy, to inhale the air by the
mouth from one bag, and to expire it from the
nofe into another. This method was purfued
with lefs than three quarts, but the effects were
fo powerful as to oblige me to take in a little
common air occafionally. I foon found my
nervous fyftem agitated by the higheft fenfations of pleafure, which are difficult of defcription ; my mufcular powers were very much
increafed, and I went on breathing with great
vehemence, not from a difficulty of infpiration,
but from an eager avidity for more air. When
the bags were exhaufted and taken from me,
I continued breathing with the fame violence,
then fuddenly ftarting from the chair, and vociferating with pleafure, I made towards thofe
that were prefent, as I wifhed they fhould
participate in my feelings. I ftruck gently at

Mr. Davy and a ftranger entering the room at the moment, I made towards him, and gave him feveral blows, but more in the fpirit of good humour than of anger. I then ran through different rooms in the houfe, and at laft returned to the laboratory fomewhat more compofed; my fpirits continued much elevated for fome hours after the experiment, and I felt no con-fequent depreffion either in the evening or the day following, but flept as foundly as ufual.

On the 5th of May, I again attempted to breathe nitrous oxide, but it happened to contain fufpended nitrous vapour which rendered it non-refpirable.

On the 7th, I infpired 7 quarts of pure gas mingled with an equal quantity of common air, the fenfations were pleafant, and my mufcular power much increafed.

On the 8th, I infpired five quarts without any mixture of common air, but the effects were not equal to thofe produced the day before; Indeed there were reafons for fuppofing that the gas was impure.

On the 18th, I breathed nearly fix quarts of the pure nitrous oxide. It is not eafy to defcribe my fenfations; they were fuperior to any thing I ever before experienced. My ftep was firm, and all my mufcular powers increafed. My fenfes were more alive to every furrounding impreffion; I threw myfelf into feveral theatrical attitudes, and traverfed the laboratory with a quick ftep ; my mind was elevated to a moft fublime height. It is giving but a faint idea of the feelings to fay, that they refembled thofe produced by a reprefentation of an heroic fcene on the ftage, or by reading a fublime paffage in poetry when circumftances contribute to awaken the fineft fympathies of the foul. In a few minutes the ufual ftate of mind returned. I continued in good fpirits for the reft of the day, and flept foundly.

Since the 18th of May, I have very often breathed nitrous oxide. In the firft experiments when pure, its effects were generally fimilar to thofe juft defcribed.

Lately I have feldom experienced vivid fen-

fations. The pleafure produced by it is flight
and tranquil, I rarely feel fublime emotions or
increafed mufcular power.

J. W. Tobin.

October, 1799.

II. *Detail of* Mr. Wm. Clayfield.

The firft time that I breathed the nitrous
oxide, it produced feelings analogous to thofe of
intoxication. I was for fome time unconfcious
of exiftence, but at no period of the experiment
experienced agreeable fenfations, a momentary
naufea followed it ; but unconnected with languor or head-ache.

After this I feveral times refpired the gas, but
on account of the fulnefs in the head and apparent throbbing of the arteries in the brain,*always
defifted to breathe before the full effects were
produced. In two experiments however, when
by powerful voluntary efforts I fucceeded in
breathing a large quantity of gas for fome mi-

* In fome of thefe experiments, hearing was rendered
more acute.

III. *Letter from* DR. KINGLAKE.

In compliance with your defire, I will endeavour to give you a faithful detail of the effects produced on my fenfations by the inhalation of nitrous oxide.

My firft infpiration of it was limited to four quarts, diluted with an equal quantity of atmofpheric air. After a few infpirations, a fenfe of additional freedom and power (call it energy if you pleafe) agreeably pervaded the region of the lungs; this was quickly fucceeded by an almoft delirious but highly pleafurable fenfation in the brain, which was foon diffufed over the whole frame, imparting to the mufcular power at once an encreafed difpofition and tone for action; but the mental effect of the excitement was fuch as to abforb in a fort of intoxicating placidity, and delight, volition, or rather the power of voluntary motion. Thefe effects were in a greater or lefs degree protracted during about five minutes, when the former ftate returned, with the difference however of feeling more cheerful and alert, for feveral hours after.

It feemed alfo to have had the further effect of reviving rheumatic irritations in the fhoulder and knee-joints, which had not been previoufly felt for many months. No perceptible change was induced in the pulfe either at or fubfequent to the time of inhaling the gas.

The effects produced by a fecond trial of its powers, were more extenfive, and concentrated on the brain. In this inflance, nearly fix quarts undiluted, were accurately and fully inhaled. As on the former occafion, it immediately proved agreeably refpirable, but before the whole quantity was quite exhaufted, its agency was exerted fo ftrongly on the brain, as progreffively to fufpend the fenfes of feeing, hearing, feeling, and ultimately the power of volition itfelf. At this period, the pulfe was much augmented (both in force and frequency ; flight convulfive twitches of the mufcles of the arms were alfo. induced ; no. painful fenfation, naufea, or languor, however, either preceded, accompanied, or followed this ftate, nor did a minute elapfe before the brain rallied, and refumed its wonted faculties, when a fenfe of glowing warmth extending over the fyftem, was fpeedily fucceeded by a re-inftatement of the equilibrium of health.

The more permanent effects were (as in the firft experiment) an invigorated feel of vital power, improved fpirits, tranfient irritations in different parts, but not fo characteriftically rheumatic as in the former inftance.

Among the circumftances moft worthy of regard in confidering the properties and adminiftration of this powerful aërial agent, may be ranked, the fact of its being (contrary to the prevailing opinion*) both highly refpirable, and

* Dr. Mitchill (an American Chemift) has erroneoufly fuppofed its full admiffion to the lungs, in its concentrated ftate, to be incompatible with animal life, and that in a more diluted form it operates as a principal agent in the

falutary, that it impreffes the brain and fyftem at large with a more or lefs ftrong and durable degree of pleafurable fenfation, that unlike the effect of other violently exciting agents, no fenfible exhauftion or diminution of vital power accrues from the exertions of its ftimulant property, that its moft exceffive operation even, is neither permanently nor tranfiently debilitating; and finally, that it fairly promifes under judicious application, to prove an extremely efficient remedy, as well in the vaft tribe of difeafes originating from deficient irritability and fenfibility, as in thofe proceeding from morbid affociations, and modifications, of thofe vital principles.

production of contagious difeafes, &c. This gratuitous pofition is thus unqualifiedly affirmed. " If a full infpira-" tion of gafeous oxyd be made, there will be a fudden " extinction of life ; and this accordingly accounts for the " fact related by Ruffel (Hiftory of Aleppo, p. 232.) and " confirmed by other obfervers, of many perfons falling " down dead fuddenly, when ftruck with the contagion of " the plague."

Vide Remarks on the Gafeous Oxyd of Azote, by Samuel Latham Mitchill, M. D.

If you ſhould deem any thing contained in
this curſory narrative capable of ſubſerving in
any degree the practical advantages likely to
reſult from your ſcientific and valuable inveſtigation of the genuine properties of the nitrous
oxide, it is perfectly at your diſpoſal.

I am

Your ſincere friend,

ROBERT KINGLAKE.

Briſtol, June 14*th,* 1799.

To MR. DAVY.

IV. *Detail of* MR. SOUTHEY.

In breathing the nitrous oxide, I could not
diſtinguiſh between the firſt feelings it occaſioned
and an apprehenſion of which I was unable to
diveſt myſelf. My firſt definite ſenſation was a
dizzineſs, a fulneſs in the head, ſuch as to induce a fear of falling. This was momentary.
When I took the bag from my mouth, I immediately laughed. The laugh was involuntary

but highly pleafurable, accompanied by a thrill all through me ; and a tingling in my toes and fingers, a fenfation perfectly new and delightful. I felt a fulnefs in my cheft afterwards ; and during the remainder of the day, imagined that my tafte and hearing were more than commonly quick. Certain I am that I felt myfelf more than ufually ftrong and chearful.

In a fecond trial, by continuing the inhalation longer, I felt a thrill in my teeth ; and breathing ftill longer the third time, became fo full of ftrength as to be compelled to exercife my arms and feet.

Now after an interval of fome months, during which my health has been materially impaired, the nitrous oxide produces an effect upon me totally different. Half the quantity affects me, and its operation is more violent ; a flight laughter is firft induced,* and a defire to continue the

---

* In the former experiments, Mr. Southey generally refpired fix quarts, now he is unable to confume two.

In an experiment made fince this paper was drawn up, the effect was rather pleafurable.

inhalation, which is counteracted by fear from
the rapidity of refpiration ; indeed my breath
becomes fo fhort and quick, that I have no doubt
but the quantity which I formerly breathed,
would now deftroy me. The fenfation is not
painful, neither is it in the flighteft degree
pleafurable.

-- ROBERT SOUTHEY.

V. *Letter from* DR. ROGET.

The effect of the firft infpirations of the nitrous oxide was that of making me vertiginous,
and producing a tingling fenfation in my hands
and feet : as thefe feelings increafed, I feemed
to lofe the fenfe of my own weight, and imagined
I was finking into the ground. I then felt a
drowfinefs gradually fteal upon me, and a dif-
inclination to motion ; even the actions of
infpiring and expiring were not performed
without effort : and it alfo required fome attention of mind to keep my noftrils clofed with my
fingers. I was gradually roufed from this tor-

por by a kind of delirium, which came on fo
rapidly that the air-bag dropt from my hands.
This sensation increased for about a minute
after I had ceased to breathe, to a much greater
degree than before, and I suddenly loft fight of
all the, objects around me, they being apparently obscured by clouds, in which were many
luminous points, similar to what is often experienced on rifing suddenly and stretching out
the arms, after fitting long in one position.

I felt myself totally incapable of speaking,
and for some time loft all consciousness of where
I was, or who was near me. My whole frame
felt as if violently agitated : I thought I panted
violently : my heart feemed to palpitate, and
every artery to throb with violence; I felt a
finging in my ears; all the vital motions
feemed to be irresistibly hurried on, as if their
equilibrium had been destroyed, and every
thing was running headlong into confusion.
My ideas succeeded one another with extreme
rapidity, thoughts rushed like a torrent through
my mind, as if their velocity had been suddenly

accelerated by the burfting of a barrier which had before retained them in their natural and equable courfe. This ftate of extreme hurry, agitation, and tumult, was but tranfient. Every unnatural fenfation gradually fubfided; and in about a quarter of an hour after I had ceafed to breathe the gas, I was nearly in the fame ftate in which I had been at the commencement of the experiment.

I cannot remember that I experienced the leaft pleafure from any of thefe fenfations. I can however, eafily conceive, that by frequent repetition I might reconcile myfelf to them, and poffibly even receive pleafure from the fame fenfations which were then unpleafant.

I am fenfible that the account I have been able to give of my feelings is very imperfect. For however calculated their violence and novelty were to leave a lafting impreffion on the memory, thefe circumftances were for that very reafon unfavourable to accuracy of comparifon with fenfations already familiar.

The nature of the fenfations themfelves,

which bore greater refemblance to a half delirious dream than to any diftinct ftate of mind
capable of being accurately remembered, con-
tributes very much to increafe the difficulty.
And as it is above two months fince I made the
experiment, many of the minuter circumftances have probably efcaped me.

I remain

Yours, &c.

P. ROGET.

To Mr. Davy.

VI. *Letter from* Mr. James Thomson.

The firft time I refpired nitrous oxide, the
experiment was made under a ftrong impreffion
of fear, and the quantity I breathed not fufficient, as you informed me, to produce the
ufual effect. I did not note very accurately my
fenfations. I remember I experienced a flight
degree of vertigo after the third or fourth
infpiration ; and breathed with increafed vigor,
my infpirations being much deeper and more

vehement than ordinary. I was enabled the next time I made the experiment, to attend more accurately to my fenfations, and you have the obfervations I made on them at the time.

After the fourth infpiration, I experienced the fame increafed action of the lungs, as in the former cafe. My infpirations became un-commonly full and ftrong, attended with a thrilling fenfation about the cheft, highly plea-furable, which increafed to fuch a degree as to induce a fit of involuntary laughter, which I in vain endeavoured to reprefs. I felt a flight giddinefs which lafted for a few moments only. My infpirations now became more vehement and frequent ; and I inhaled the air with an avidity ftrongly indicative of the pleafure I received. That peculiar thrill which I had at firft experienced at the cheft, now pervaded my whole frame ; and during the two or three laft infpirations, was attended with a remarkable tingling in my fingers and toes. My feelings at this moment are not to be defcribed : I felt a high, an extraordinary degree of pleafure, different from that produced by wine, being diveſted of all its groſs accompaniments, and yet approaching nearêr to it than to any other fenſation. I am acquainted with.

I am certain that my muſcular ſtrength was for a time much increaſed. My diſpoſition to exert it was ſuch as I could not repreſs, and the ſatisfaction I felt in any violent exertion of my legs and arms is hardly to be conceived. Theſe vivid fenſations were not of long duration ; they diminiſhed infenſibly, and in little more than a quarter of an hour I could perceive no difference between the ſtate I was then in, and that previous to the reſpiration of the air.

The obſervations I made on repeating the experiment, do not differ from the preceding, except in the circumſtance of the involuntary laughter, which I never afterwards experienced, though I breathed the air feveral times ; and in the following curious fact, which, as it was dependent on circumſtances, did not always occur.

Having reſpired the fame quantity of air as ufual, and with preciſely the fame effects, I was furprifed to find myfelf affected a few minutes afterwards with the recurrence of a pain in my back and knees, which I had experienced the preceding day from fatigue in walking. I was rather inclined to deem this an accidental coincidence than an effect of the air; but the fame thing conftantly occurring whenever I breathed the air, fhortly after fuffering pain either from fatigue, or any other accidental caufe, left no doubt on my mind as to the accuracy of the obfervation.

I have now given you the fubftance of the notes I made whilft the impreffions were ftrong on my mind. I cannot add any thing from recollection that will at all add to the accuracy of this account, or affift thofe who have not refpired this air, in forming a clearer idea of its extraordinary effects. It is extremely difficult to convey to others by means of words, any idea of particular fenfations, of which they have had no experience. It can only be done by making ufe of fuch terms as are expreffive of fenfations that refemble them, and in thefe our vocabulary is very defective. To be able at all to comprehend the effects of nitrous oxide, it is neceffary to refpire it, and after that, we muft either invent new terms to exprefs thefe new and particular fenfations, or attach new ideas to old ones, before we can communicate intelligibly with each other on the operation of this extraordinary gas.

I am &c.

JAMES THOMSON.

London, Sept. 21, 1799.

To MR. DAVY.

VII. *Detail of* MR. COLERIDGE.

The firft time I infpired the nitrous oxide, I felt an highly pleafurable fenfation of warmth over my whole frame, refembling that which I remember once to have experienced after returning from a walk in the fnow into a warm room. The only motion which I felt inclined to make, was that of laughing at thofe who were looking at me. My eyes felt diftended,

and towards the laſt, my heart beat as if it were
leaping up and down. On removing the mouth-
piece the whole fenſation went off almoſt
inſtantly.

The fecond time, I felt the fame pleaſurable
fenſation of warmth, but not I think, in quite
fo great a degree. I wiſhed to know what effect
it would have on my impreſſions; I fixed my
eye on fome trees in the diſtance, but I did
not find any other effect except that they became dimmer and dimmer, and looked at laſt
as if I had feen them through tears. My heart
beat more violently than the firſt time. This
was after a hearty dinner.

The third time I was more violently acted
on than in the two former. Towards the laſt,
I could not avoid, nor indeed felt any wiſh to
avoid, beating the ground with my feet; and
after the mouth-piece was removed, I remained
for a few feconds motionlefs, in great extacy.

The fourth time was immediately after break-
faſt. The few firſt inſpirations affected me fo
little that I thought Mr. Davy had given me

atmofpheric air : but foon felt the warmth beginning about my cheft, and fpreading upward
and downward, fo that I could feel its progrefs
over my whole frame. My heart did not beat
fo violently ; my fenfations were highly pleafurable, not fo intenfe or apparently local, but
of more unmingled pleafure than I had ever
before experienced.*

S. T. COLERIDGE.

VIII. *Detail of* MR. WEDGWOOD.

July 23, I called on Mr. Davy at the Medical Inftitution, who afked me to breathe fome
of the nitrous oxide, to which I confented,
being rather a fceptic as to its effects, never
having feen any perfon affected. I firft breathed
about fix quarts of air which proved to be only
common atmofpheric air, and which confequently produced no effect.

I then had 6 quarts of the oxide given me in

---

* The dofes in thefe experiments were from five to
feven quarts.

a bag undiluted, and as foon as I had breathed
three or four refpirations, I felt myfelf affected
and my refpiration hurried, which effect increafed rapidly until I became as it were entranced,
when I threw the bag from me and kept breathing on furioufly with an open mouth and holding my nofe with my left hand, having no
power to take it away though aware of the
ridiculoufnefs of my fituation. Though apparently deprived of all voluntary motion, I was
fenfible of all that paffed, and heard every thing
that was faid ; but the moft fingular fenfation
I had, I feel it impoffible accurately to defcribe.
It was as if all the mufcles of the body were
put into a violent vibratory motion ; I had a very
ftrong inclination to make odd antic motions
with my hands and feet. When the firft ftrong
fenfations went off, I felt as if I were lighter than
the atmofphere, and as if I was going to mount
to the top of the room. I had a metallic tafte
left in my mouth, which foon went off.

Before I breathed the air, I felt a good deal
fatigued from a very long ride I had had the

day before, but after breathing, I loft all fenfe
of fatigue.

IX. *Detail of* Mr. George Burnet,

I had never heard of the effects of the nitrous
oxide, when I breathed fix quarts of it. I felt
a delicious tremor of nerve, which was rapidly
propagated over the whole nervous fyftem. As
the action of inhaling proceeds, an irrefiftible
appetite to repeat it is excited. There is now a
general fwell of fenfations, vivid, ftrong, and
inconceivably pleafurable. They ftill become
more vigorous and glowing till they are communicated to the brain, when an ardent flufh
overfpreads the face. At this moment the tube
inferted in the air-bag was taken from my
mouth, or I muft have fainted in extacy.

. The operation being over, the ftrength and
turbulence of my fenfations fubfided. To this
fucceeded a ftate of feeling uncommonly ferene
and tranquil. Every nerve being gently agita-

ted with a lively enjoyment. It was natural to expect that the effect of this experiment, would eventually prove debilitating. So far from this I continued in a state of high excitement the remainder of the day after two o'clock, the time of the experiment, and experienced a flow of spirits not merely chearful, but unusually joyous.

GEORGE BURNET.

X. *Detail of* MR. T. POPLE.

A disagreeable sensation as if breaking out into a profuse perspiration, tension of the tympanum, cheeks, and forehead; almost total loss of muscular power; afterwards increased powers both of body and mind, very vivid sensations and highly pleasurable. Those pleasant feelings were not new, they were felt, but in a less degree, on ascending some high mountains in Glamorganshire.

On taking it the second time, there was a disagreeable feeling about the face. In a few feconds; "the feelings became pleafurable ; all the faculties abforbed by the fine pleafing feelings of exiftence. without confcioufnefs ; an involuntary burft of laughter.

THOMAS POPLE.

XI. *Detail of* MR. HAMMICK.

Having never heard any thing of the mode of operation of nitrous oxide, I breathed gas in a filk bag for fome time, and found no effects, but oppreffion of refpiration. Afterwards Mr. Davy told me that I had been breathing atmofpheric air.

In a fecond experiment made without knowing what gas was in the bag; I had not breathed half a minute, when from the extreme pleafure I felt, I unconcioufly removed the bag from my mouth ; but when Mr. Davy offered to take it from me, I refufed to let him have it, and faid eagerly, " let me breathe it again, it is highly pleafant ! it is the ftrongeft ftimulant I ever felt !" I was cold when I began to refpire, but had immediately a pleafant glow extending to my toes and fingers. I experienced from the air a pleafant tafte which I can only call fweetly aftringent; it continued for fome time: the fenfe of exhilaration was lafting. This air Mr. Davy told me was nitrous oxide.

In another experiment, when I breathed a fmall dofe of nitrous oxide, the effects were flight, and fometime afterwards I felt an unufual yawning and languor.

The laft time that I breathed the gas, the feelings were the moft pleafurable I ever experienced; my head appeared light, there was a great warmth in the back and a general unufual glow; the tafte was diftinguifhable for fome time as in the former experiment. My ideas were more vivid, and followed the natural order of affociation. I could not refrain from mufcular action.

STEPHEN HAMMICK, Junr.

Sept. 15th.

XII. *Detail of* Dr. Blake.

Dr. Blake inhaled about fix quarts of the air, was affected during the procefs of refpiring it with a flight degree of vertigo, which was almoft immediately fucceeded by a thrilling fenfation extending even to the extremities, accompanied by a moft happy ftate of mind and highly pleafurable ideas. He felt a great propenfity to laugh, and his behaviour in fome meafure appeared ludicrous to thofe around him. Mufcular power feemed agreeably increafed, the pulfe acquired ftrength and firmnefs, but its frequency was fomewhat diminifhed. He perceived rather an unpleafant tafte in the mouth and about the fauces for fome hours afterwards, but in every other refpect, his feelings were comfortable during the remainder of the day.

December, 30*th.*

To Mr. Davy.

XIII. *Detail of* Mr. Wansey.

I breathed the gas out of a filk bag, believing it to be nitrous oxide, and was much furprifed to find that it produced no fenfations. After the experiment, Mr. Davy told me it was common air.

. I then breathed a mixture of common air and nitrous oxide. I felt a kind of intoxication in the middle of the experiment, and ftopping to exprefs this, deftroyed any farther effeds.

I now breathed pure nitrous oxide ; the effed was gradual, and I at firft experienced fulnefs in the head, and afterwards fenfations fo delightfut, that I can compare them to no others, except thofe which I felt (being a lover of mufic) about five years fince in Weftminfter Abbey, in fome of the grand choruffes in the Meffiah, from the united powers of 700 inftruments of mufic. I continued exhilarated throughout the day, flept at night remarkably found, and experienced when I awoke in the morning, a recurrence of pleafing fenfation.

In another experiment, the effect was ſtill greater, the pulſe was rendered fuller and quicker, I felt a fenfe of throbbing in the head with highly pleafurable thrillings all over the frame. The new feelings were at laſt fo powerful as to abforb all perception. I diſtinguiſhed during and after the experiment, a taſte on the tongue, like that produced by the contact of zinc and filver.

HENRY WANSEY.

XIV. *Detail of* MR. RICKMAN.

On inhaling about ſix quarts, the firſt altered feeling was a tingling in the elbows not unlike the effect of a flight electric ſhock. Soon afterwards, an involuntary and provoking dizzineſs as in drunkenneſs. Towards the cloſe of the inhalation, this ſymptom decreaſed; though the noſe was ſtill involuntary held faſt after the air-bag was removed. The doſe was probably an undercharge, as no extraordinary ſenſation was felt more than half a minute after the inhalation.

J. RICKMAN.

XV. *Detail of* MR. LOVELL EDGWORTH.

My firſt ſenſation was an univerſal and conſiderable tremor. I then perceived ſome giddineſs in my head, and a violent dizzineſs in my ſight; thoſe ſenſations by degrees ſubſided, and I felt a great propenſity to bite through the wooden mouth-piece, or the tube of the bag through which I inſpired the air. After I had breathed all the air that was in the bag, I eagerly wiſhed for more. I then felt a ſtrong propenſity to laugh, and did burſt into a violent fit of laughter, and capered about the room without having the power of reſtraining myſelf. By degrees theſe feelings ſubſided, except the tremor which laſted for an hour after I had breathed the air, and I felt a weakneſs in my knees. The principal feeling through the whole of the time, or what I ſhould call the characteriftical part of the effect, was a total difficulty of reftraining my feelings, both corporeal and mental, or in other words, not having any command of one's felf.

XVI. *Detail of* MR. G. BEDFORD.

I inhaled 6 quarts. Experienced a fenfation of fulnefs in the extremities and in the face, with a defire and power of expanfion of the lungs very pleafurable. Feelings fimilar to intoxication were produced, without being difagreeable. When the bag was taken away, an involuntary though agreeable laughter took place, and the extremities were warm.

In about a quarter of an hour after the above experiment, I inhaled 8 quarts. The warmth and fulnefs of the face and extremities were fooner produced during the infpiration. The candle and the perfons about me, affumed the fame appearances as took place during the effect produced by wine, and I could perceive no determinate outline. The defire and power to expand the lungs was increafed beyond that in the former experiment, and the whole body and limbs feemed dilated without the fenfe of tenfion, it was as if the bulk was increafed without any addition to the fpecific gravity of the body, which was highly pleafant. The provocation to laughter was not fo great as in the former experiment, and when the bag was removed, the warmth almoft fuddenly gave place to a coldnefs of the extremities, particularly of the hands which were the firft to become warm during the infpiration. A flight fenfation of fulnefs not amounting to pain in the head, has continued for fome minutes. After the firft experiment, a fenfation in the wrifts and elbows took place, fimilar to that produced by the electric fhock.

G. C. BEDFORD.

March 30*th*, 1800.

XVII. *Detail of* Miss Ryland.

After having breathed five quarts of gas, I experienced for a fhort time a quicknefs and difficulty of breathing, which was fucceeded by extreme languor, refembling fainting, without the very unpleafant fenfation with which it is ufually attended. It entirely deprived me of the power of fpeaking, but not of recollection, for I heard every thing that was faid in the room during the time; and Mr. Davy's remark "that my pulfe was very quick and full." When the languor began to fubfide, it was fucceeded by reftlefsnefs, accompanied by involuntary mufcular motions. I was warmer than ufual, and very fleepy for feveral hours.

XVIII. *Letter from* Mr. M. M. Coates.

I will, as you requeft, endeavour to defcribe to you the effect produced on me laft Sunday fe'nnight by the nitrous oxide, and will at the fame time tell you what was the previous ftate of my mind on the fubject.

When I fat down to breathe the gas, I believed that it owed much of its effect to the predifpo-fing agency of the imagination, and had no expectation of its fenfible influence on myfelf. Having ignorantly breathed a bag of common air without any effect, my doubts then arofe to pofitive unbelief.

After a few infpirations of the nitrous oxide, I felt a fulnefs in my head, which increafed with each inhalation, until, experiencing fymp-toms which I thought indicated approaching fainting, I ceafed to breathe it, and was then confirmed in my belief of its inability to pro-duce in me any pleafurable fenfation.

But after a few feconds, I felt an immoderate flow of fpirits, and an irrefiftible propenfity to violent laughter and dancing, which, being fully confcious of the violence of my feelings, and of their irrational exhibition, I made great but ineffectual efforts to reftrain ; this was my ftate for feveral minutes. During the reft of the day,

I experienced a degree of hilarity altogether new to me. For fix or feven days afterwards, I feemed to feel moft exquifitely at every nerve, and was much indifpofed to my fedentary purfuits; this acute fenfibility has been gradually diminifhing; but I ftill feel fomewhat of the effects of this novel agent.

Your's truly,

To Mr. DAVY.   M. M. COATES.

June 11th, 1800.

DIVISION III.

I. *Abstracts from additional Details.*

THE trials related in the following abftracts, have been chiefly made fince the publication of Dr. Beddoes's Notice. Many of the individuals breathed the gas from pure curiofity. Others with a difbelief of its powers.

MR. WYNNE, M. P. breathed five quarts of diluted nitrous oxide, without any fenfation. Six quarts produced fulnefs in the cheft, heat in the hands and feet, and fenfe of tenfion in the fingers, flight but pleafant fenfations. Seven quarts produced no new or different effects.

Mr. MACKINTOSH feveral times breathed nitrous oxide. He had fenfe of fulnefs in the head, thrillings, tingling in the fingers, and generally pleafurable feelings.

Mr. JOHN CAVE, Junr. from breathing four fuperior wine, and general pleafant feelings.

Mr. MICHAEL CASTLE, from five quarts, experienced fenfations of heat and thrilling, general fpirits heightened confiderably as from wine; afterwards, flight pain in the back of the head.

Mr. H. CARDWELL, from five quarts, had feelings fo pleafurable as almoft to deftroy confcioufnefs; almoft convulfed with laughter; for a long time could not think of the feeling without laughing; fenfation of lightnefs for fome time after.

Mr. JARMAN, from five quarts, great pleafure, laughter, certainly better fpirits, glow in the cheeks which continued long.

The gentleman who furnifhed the preceding detail, had heard of the effects of nitrous oxide, and was prepared to experience new fenfations: I therefore gave him a bag of common air, which he refpired, believing it to be nitrous oxide; and was much furprifed that no effects were produced. He then breathed five quarts of nitrous oxide, and after the experiment, gave this account of his fenfations.

Rev. W. A. CANE, after inhaling the gas, felt the moft delicious fenfations accompanied by a thrill through every part of his body. He did not think it poffible fo charming an effect could have been produced. He had heard of the gas; but the refult of the experiment far exceeded his expectations.

May 6th, 1800.

Mr. JOSEPH PRIESTLEY from breathing nitrous oxide, generally had unpleafant fulnefs of the head and throbbing of the arteries, which prevented him from continuing the refpiration.

Dr. Beddoes mentioned in his Notice, that Mr. JOSIAH WEDGWOOD and Mr. THOMAS WEDGWOOD, experienced rather unpleasant feelings from the gas. Mr. JOSIAH WEDGWOOD has since repeated the trial, the effects were powerful, but not in the slightest degree pleasant.

Mr. R. BOULTON and Mr. G. WATT have been much less affected than any individuals.

Many other persons have respired the gas, but as their accounts contain nothing unnoticed in the details, it is useless to particularise them.

The cases of all the males who have been unpleasantly affected since we have learnt to prepare the gas with accuracy, are related in this Section and in the last Division. Those who have been pleasurably affected after a fair trial and whose cases are not noticed, generally experienced fulness in the head, heat in the chest, pleasurable thrillings, and consequent exhilaration.

To persons who have been unaccustomed to breathe through a tube, we have usually given common air till they have learnt to respire with accuracy: and in cafes where the form of the mouth has prevented the lips from being accurately clofed on the breathing tube, by the advice of Mr. Watt, we have ufed a tin plate conical mouth-piece fixed to the cheeks, and accurately adapted to the lips ; by means of which precautions, all our later trials have been perfectly conclufive.

II. *Of the effects of Nitrous Oxide upon perfons inclined to hyfterical and nervous affections.*

The cafe of Mifs —— N. and other cafes, detailed by Dr. Beddoes in his Notice, feemed to prove that the action of nitrous oxide was capable of producing hyfterical and nervous affections in delicate and irritable confti-tutions.

On this fubject, we have lately acquired additional facts.

Mifs E. a young lady who had been fubject to hyfteric fits, breathed three quarts of nitrous oxide mingled with much common air, and felt no effects but a flight tendency to fainting.; She then breathed four quarts of pure nitrous, oxide : her firft infpirations were deep,. her, laft very feeble. . At the end fhe dropt the bag from her lips, and continued for fome moments motionlefs. Her pulfe which at the beginning of the experiment was ftrong, appeared to me to be at this time, quicker and weaker. She foon began to move her hands and talked . for, fome minutes incoherently, as if ignorant of. what had paffed. In lefs than a quarter of an hour, fhe had recovered, but could give no account of her fenfations. A certain degree of languor continued through the day.;

A young lady who never had hyfterical attacks, wifhed to breathe the gas. I informed her of the difagreeable effects it had fome-times produced, and advifed her if fhe had the flighteft tendency to nervous affection, not to make the trial. She perfifted in her refolution.

To afcertain the influence of imagination,

I firſt gave her a bag of common air, which ſhe declared produced no effect. I then ordered for her a quart of nitrous oxide mingled with two quarts of common air; but from the miſtake of the perſon who prepared it, three quarts of nitrous oxide were adminiſtered with one of common air. She breathed this for near a minute, and after the experiment, deſcribed her ſenſations as unpleaſant, and ſaid ſhe felt at the moment as if ſhe was dying. The unpleaſant feelings quickly went off, and a few minutes after, ſhe had apparently recovered her former ſtate of mind. In the courſe of the day, however, a violent head-ache came on, and in the evening after ſhe had taken a medicine which operated violently, hyſterical affections were produced, followed by great debility. They occaſionally returned for many days, and ſhe continued weak and debilitated for a great length of time.

Mrs. S. a delicate lady, liable to nervous affections who had heard of the caſes juſt detailed, chofe to breathe the gas. By three quarts fhe was thrown into a trance, which lafted for three or four minutes. On recovering, fhe could give no accòunt of her feelings, and had fome languor for half an hour afterwards,

Thefe phænomena have rendered us cautious in adminiftering the gas to delicate females. In a few inftances however, it has been taken by perfons of this clafs, and even by thofe inclined to hyfterical and nervous complaints with pleafurable effects,

Mifs L. a young lady who had formerly had hyfterical fits, breathed a quart of nitrous oxide with three quarts of common air without effects. Two quarts of nitrous oxide with one of common air produced a flight giddinefs; four quarts of nitrous oxide produced a fit of immoderate laughter, which was fucceeded by flight exhilaration, her fpirits were good throughout the day, and no depreffion followed.

Mifs B. Y—— and Mifs S. Y—— both delicate but healthy young ladies, were affected very pleafantly ; each by three quarts of nitrous oxide, the firft time of refpiring it. Mifs B Y——continued exhilarated and in high fpirits for fome hours after the dofe. Mifs S. Y— had a flight head-ache, which did not go off for fome hours.

Mrs. F. inclined to be hyfterical, breathed four quarts of nitrous oxide mingled with common air. She was giddy and defcribed her feelings as odd ; but had not the flighteft languor after the experiment.

III. *Obfervations on the effects of Nitrous Oxide, by* DR. BEDDOES.

Neither my notes nor my recollection fupply much in addition to what I formerly ftated in the *Notice of Obfervations at the Pneumatic Inflitution. Longman.* The gas maintains its firft character as well in its effects on me, as in

the benefit it confers on fome of the paralytic,
and the injury it does or threatens to the hyfterical and the exquifitely fenfible. I find that
five or fix quarts operate as powerfully as ever.
I feem to make a given quantity go farther by
holding my breath fo that the gas may be abforbed in a great degree without returning into
the bag, and therefore, be as little heated before infpiration as poffible.—This may be fancy.

After innumerable trials, I have never once
felt laffitude or depreffion* Moft commonly

---

* Of the facts on which Brown founded his law of indirect debility, no prudent man will lofe fight either in
practifing or ftudying medicine. They are incontroverti-
ble.—And our new facts may doubtlefs be conciliated to
the Brunonian doctrine.

But to fuppofe that the expenditure of a quality or a fubftance or a fpirit, and its renewal or accumulation are the
general principles of animal phænomena, feems to me a griev-
ous and baneful error. I believe it often happens that excitement and excitability increafe, and that they oftener decreafe
together;—In fhort, without generalizing in a manner, of
which Brown and fimilar theorifts had no conception, our
notions of the living world will in my opinion, continue
to be as confufed as the elements are faid to have been in
chaos. On fome future occafion, I may prefume to point

I am fenfible of a grateful glow *circum præcor-*
dia. This has continued for hours.—In two or
three inftances only has exhalation failed to be
followed by pleafurable feeling, it has never been
followed by the contrary. On a few occafions
before the gas was exhaufted, I have found it
impóffible to continue breathing.

The pulfe at firft becomes fuller and ftronger.
Whenever, after expofure to a cold wind, the
warmth of the room has created a glow in the
cheeks, the gas has increafed this to ftrong
flufhing—which common air breathed in the
fame way, failed to do.

Several times I have found that a cut which
had ceafed to be painful has fmarted afrefh, and
on taking two dofes in fucceffion, the fmarting
ceafed in the interval and returned during the
fecond refpiration. I had no previous expectation of the firft fmarting.

out the region through which I imagine the path to wind,
that will lead the obfervers of fome diftant generation to a
point, whence they may enjoy a view of the fubtle, bufy
and intricate movements of the organic creation as clear as
Newton obtained of the movements of the heavenly maffes.

The only time I was near rendering myfelf infenfible to prefent objects by very carefully breathing feveral dofes in quick fucceffion, I forcibly exclaimed, TONES !—In fact, befides a general thrilling, there feemed to be quick and ftrong alterations in the degree of illumination of all furrounding objects ; and I felt as if compofed of finely vibrating ftrings. On this occafion, the fkin feemed in a ftate of conftriction and the lips glued to the mouth-piece, and the mucous membrane of the lungs contracted, but not painfully. However, no conftriction or corrugation of the fkin could be feen. I am confcious of having made a great number of obfervations while breathing, which I could never recover.

Immediately afterwards I have often caught myfelf walking with a hurried ftep and bufy in foliloquy. The condition of general fenfation being as while hearing chearful mufic, or after good news, or a moderate quantity of wine.

Mr. John Cave, Junr. and his three friends, as well as others, compared the effects to Champagnè. Moft perfons have had the idea of the effect of fermented 'liquors excited by the gas. It were to be wifhed that we had, for a ftandard of comparifon, obfervations on the effect of thefe liquors as diverfified and as accurate as we have obtained concerning the gas ; nor would more uniformity in the action, of thefe fub-ftances be obferved if the enquiry were ftrictly purfued. Opium and fpirits feem, in particular, ftates to ficken, and diftrefs in the firft inftance; how differently does wine at an early hour and fafting act upon thofe who are accuftomed to take it only after dinner!

· I thought it might be an amufing fpectacle to fee the different tints of blood flowing from a wound by a leech in confequence of breathing different airs. The purple from the nitrous oxide was very evident. Oxygene, we thought, occafioned a quicker flow and brighter color in the blood. In another experiment, an inflamed area round the puncture from a leech applied the day before, was judged by feveral fpectators to become much more crimfon on the refpiration of about 20 quarts of oxygene gas, which poffibly acts more powerfully on inflamed parts.* Thefe and many fimilar experiments, require to be repeated on the blood of fingle arteries opened in warm and cold animals.

It has appeared to me that I could hold my breath uncommonly long when refpiring oxygene gas mixed with nitrous oxide. While trying this to-day, (17th June), I thought the fenfe of fmell much more acute after the nitrous oxide than before I began to refpire at all ; and then I felt confcious that this increafed acutenefs had before repeatedly occurred—a

---

* After writing this, I was prefent when an invalid, in whofe foot the gout, after much wandering, had at laft fixed, breathed 12 quarts of oxygene gas. While breathing, he eagerly pointed to the inflamed leg ; and afterwards faid he had felt in it a new fenfation, fomewhat like tenfion.—I never had feen oxygene refpired where there was fo much local inflammation.

June 18. After four quarts of oxygene with 6 of nitrous oxide and then 6 of nitrous oxide alone, violent itching of the wounds made by the leech ; and rednefs and tumour.— Both had healed, and I did not expect to feel any thing more from them.—I tried this again with two dofes of nitrous oxide—The yellow halo round one wound changed to crimfon, and there was fo much ftinging and fwelling that I feared fuppuration.—Abforption here was rapid.

fact very capable, I apprehend, of a pneumatological interpretation.

Time by my feelings has always appeared
longer than by a watch.

I thought of trying to obferve whether while
I alternately breathed quantities of nitrous oxide
and oxygene gas and common air, I could obferve any difference in the operation of a blifter
beginning to bite the fkin. It would be of
confequence to afcertain the effect of regulating
by compreffion the flow of blood, while ftimulants of various kinds (and heated bodies among
the reft) were applied to or near the extremities—becaufe in erifipelas and various inflam-
matory affections, a ready and pleafant cure
might be effected by partial compreffion of the
arteries going to the difeafed part; and a great
improvement in practice thus obtained.

But I fhould run into an endlefs digreffion,
were I to enumerate poffible phyfiological experiments with artificial airs, or to fpeculate on
the mechanical improvement of medicine, which
at prefent as far as mechanical means of affect-

ing the living fyftem are concerned, is with us in a ftate that would almoft difgrace a nation of favages.

IV. CONCLUSION.

'From the facts detailed in the preceding pages, it appears that the immediate effects of nitrous oxide upon the living fyftem, are analogous to thofe of diffufible ftimuli. Both increafe the force of circulation, produce pleafurable feeling, alter the condition of the organs of fenfation, and in their moft extenfive action deftroy life.

In the mode of operation of nitrous oxide and diffufible ftimuli, confiderable differences however, exift.

Diffufible ftimuli act immediately on the mufcular and nervous fibre. Nitrous oxide operates upon them only by producing peculiar changes in the compofition of the blood.

Diffufible ftimuli affect that part of the fyftem

moſt powerfully to which they are applied, and act on the whole only by means of its ſympathy with that part. Nitrous oxide in combination with the blood, is univerſal in its application and action. . . .

We know very little of the nature of excitement ; as however, life depends immediately on certain changes effected in the blood, in reſpiration, and ultimately, on the ſupply of certain nutritive matter by the lymphatics ; it is reaſonable to conclude, that during the action of ſtimulating ſubſtances, from the increaſed force of circulation, not only more oxygene and perhaps nitrogene muſt be combined with the blood in reſpiration,* but likewiſe more fluid nutritive matter ſupplied to it in circulation.

---

* See Dr. Beddoes's *Conſiderations, part* 1. *page* 26. His obſervations in the note in the laſt ſection, will likewiſe apply here.—Is not healthy living action dependant upon a certain equilibrium between the principles ſupplied to the blood by the pulmonary veins from reſpiration and by the lymphatics from abſorption ? Does not ſenſibility more immediately depend upon reſpiration ? Deprive an animal under ſtimulation, of air, and it inſtantly dies ; probably

By this oxygene and nutritive matter excitability may be kept up : and exhauftion confequent to excitement only produced, in confequence of a deficiency of fome of the nutritive
principles, which are fupplied by abforption.

When nitrous oxide is breathed, nitrogene
(a principle under common circumftances chiefly
carried into the blood by the abforbents in fluid
compounds) is fupplied in refpiration ; a greater
quantity of oxygene is combined with the blood
than in common refpiration, whilft lefs carbonic
acid and probably lefs water are evolved.

Hence a fmaller quantity of nutritive matter
is probably required from the abforbents during
the excitement from nitrous oxide, than during
the operation of ftimulants ; and in confequence,
exhauftion rom the expenditure of nutritive
matter more feldom occafioned.

---

if abforption could be prevented, it would likewife fpeedily
die. It would be curious to try whether intoxication from
fermented liquors cannot be prevented by breathing
during their operation, an atmofphere deprived of part of
its oxygene.

Since Refearch III. has been printed, I have endeavoured to afcertain the quantities of nitrogene produced when nitrous oxide is refpired for a confiderable time. In one experiment, when I breathed about four quarts of gas in a glafs bell over impregnated water for near a minute, it was diminifhed to about two quarts; and the refiduum extinguifhed flame.

Now the experiments in Refearch II. prove that when nitrous oxide is decompofed by combuftible bodies, the quantity of nitrogene evolved is rather greater in volume than the pre-exifting nitrous oxide. Hence much of the nitrogene taken into the fyftem during the refpiration of nitrous oxide, muft be either carried into new combinations, or given out by the capillary veffels through the fkin.

It would be curious to afcertain whether the quantity of ammoniac in the faline matters held in folution by the fecreted fluids is increafed after the refpiration of nitrous oxide. Experiments made upon the confumption of nitrous oxide mingled with atmofpheric air by the smaller animals, would go far to determine whether any nitrogene is given out through the skin.

The various effects of nitrous oxide, upon different individuals and upon the same individuals at different times, prove that its powers are capable of being modified both by the peculiar condition of organs, and by the state of general feeling.

Reasoning from common phænomena of sensation, particularly those relating to heat, it is probable that pleasurable feeling is uniformly connected with a moderate increase of nervous action; and that this increase when carried to certain limits, produces mixed emotion or sublime pleasure; and beyond those limits occasions absolute pain.

Comparing the facts in the last division, it is likely that individuals possessed of high health and little sensibility, will generally be less pleasurably affected by nitrous oxide than such as have more sensibility; in whom the emotions will sometimes so far enter the limits of pain as to become sublime ;* whilst the nervous action in such as have exquisite sensibility, will be so much increased as often to produce disagreeable feeling.

Modification of the powers of nitrous oxide by mixture of the gas with oxygene or common air, will probably enable the most delicately sensible to respire it, without danger, and even with pleasurable effects : heretofore it has been administered to such only in its pure form or mingled with small quantities of atmospheric air, and in its pure form even the most robust are unable to respire it with safety for more than five minutes.

The muscular actions† sometimes connected

* Sublime emotion with regard to natural objects, is generally produced by the connection of the pleasure of beauty with the passion of fear.

† The immortal Hartley has demonstrated that all our motions are originally automatic, and generally produced by the action of tangible things on the muscular fibre.

The common actions of adults may be distinguished into two kinds ; voluntary actions, and mixed automatic actions. The first are produced by ideas, or by ideas connected with passions. The second by impression, or by pleasure and pain.

with the feelings produced by nitrous oxide, seem
to depend in a great measure upon the particular habits of the individual; they will usually
be of that kind which is produced either by
common pleasurable feelings or strong emotions.

Hysterical affection is occasioned by nitrous oxide, probably only in consequence of the strong
emotion produced, which destroys the power of
the will, and calls up series of automatic motions
formerly connected with a variety of less powerful
but similar feelings.

The quickness of the operation of nitrous
oxide, will probably render it useful in cases of
extreme debility produced by deficiency of

In voluntary action, regular, associations of ideas and
muscular motions exist; as when a chemist performs a pre-
conceived experiment.

In mixed automatic actions, the simple motions produced
by impression are connected with series of motions formerly
voluntary, but now produced without the intervention of
ideas: as when a person accustomed to play on the harpsi-
chord, from accidentally striking a key, is induced to per-
form the series of motions which produce a well-remembered
tune.

Evidently the muscular actions produced by nitrous
oxide are mixed automatic motions.

common exciting powers. Perhaps it may be advantageoufly applied mingled with oxygene or common air, to the recovery of perfons apparently dead from fuffocation by drowing or hanging.

The only difeafes in which nitrous oxide has been hitherto employed, are thofe of defficient fenfibility.—An account of its agency in paralytic affections, will be fpeedily publifhed by Dr. Beddoes.

As by its immediate operation the tone of the irritable fibre is increafed, and as exhauftion rarely follows the violent mufcular motions fometimes produced by it, it is not unreafonable to expect advantages from it in cafes of fimple mufcular debility.

The apparent general tranfiency of its operation in the pure form in fingle dofes has been confidered as offering arguments againft its power of producing lafting changes in the conftitution. It will, however, be eafy to keep up excitement of different degrees of intenfity for a great length of time, either by adminiftering

the unmingled gas in rapid fucceffive dofes, or.
by preferving a permanent atmofphere, containing different proportions of nitrous oxide
and common air, by means of a breathing chamber.* That fingle dofes neverthelefs, are capable
of producing permanent effects in fome confti-
tutions, is evident, as well from the hyfterical
cafes as from fome of the details—particularly
that of Mr. M. M. Coates.

As nitrous oxide in its extenfive operation
appears capable of deftroying phyfical pain, it
may probably be ufed with advantage during
furgical operations in which no great effufion
of blood takes place.

From the ftrong inclination of thofe who have
been pleafantly affected by the gas to refpire it
again, it is evident, that the pleafure produced,
is not loft, but that it mingles with the mafs of
feelings, and becomes intellectual pleafure, or
hope. The defire of fome individuals acquainted
with the pleafures of nitrous oxide for the gas
has been often fo ftrong as to induce them to

* See R. IV. Div. I. page 478.

breathe with eagernefs, the air remaining in the
bags after the refpiration of others.

As hydrocarbonate acts as a fedative,† and
diminifhes living action as rapidly as nitrous
oxide increafes it, on the common theory of
excitability‡ it would follow, that by differently
modifying the atmofphere by means of this gas
and nitrous oxide, we fhould be in poffeffion
of a regular feries of exciting and depreffing*
powers applicable to every deviation of the conftitution from health : but the common theory

† R. IV. Div. I. page 467.

‡ That of Brown modified by his difciples.

* Suppofing the increafe or diminution of living action
when produced by different agents, uniform, fimilar and
differing only in degree; it would follow, that certain
mixtures of hydrocarbonate and nitrous oxide, or hydrogene and nitrous oxide, ought to be capable of fupporting
the life of animals for a much longer time than pure nitrous
oxide. From the experiments in Ref. III. Div. I. it appears
however, that this is not the cafe.
It would feem, that in life, a variety of different corpufcular changes are capable of producing phænomena
apparently fimilar ; fo that in the fcience of living action,
we are incapable of reafoning concerning caufes from effects.

of excitability is moſt probably founded on a falſe generaliſation. The modifications of diſeaſed action may be infinite and ſpecific in different organs; and hence out of the power of agents operating on the whole of the ſyſtem.

Whenever we attempt to combine our ſcattered phyſiological facts, we are ſtopped by the want of numerous intermediate analogies; and ſo looſely connected or ſo independant of each other, are the different ſeries of phænomena, that we are rarely able to make probable conjectures, much leſs certain predictions concerning the reſults of new experiments.

An immenſe maſs of pneumatological, chemical, and medical information muſt be collected, before we ſhall be able to operate with certainty, on the human conſtitution.

Pneumatic chemiſtry in its application to medicine, is an art in infancy, weak, almoſt uſeleſs, but apparently poſſeſſed of capabilities of improvement. To be rendered ſtrong and mature, ſhe muſt be nouriſhed by facts, ftrengthened by exercife, and cautioufly directed in the application of her powers by rational fcepticifm.

APPENDIX.

No. I.

Effects of Nitrous Oxide on Vegetation.

In July 1799, I introduced two small plants of spurge into nitrous oxide, in contact with a little water over mercury; after remaining in it two days, they preserved their healthy appearance, and I could not perceive that any gas had been absorbed. I was prevented by an accident, from keeping them longer in the gas.

A small plant of mint introduced into nitrous oxide and exposed to light, in three days became dark olive and spotted with brown; and in about six days was quite dead.—Another similar plant, kept in the dark in nitrous oxide,

M m

did not alter in color for five days, and at the end of feven days, was only a little yellower, than before. I could not afcertain whether any gas had been abforbed.

I introduced into nitrous oxide through water, a healthy budding rofe, thinking that its colors might be rendered brighter by the gas. I was difappointed, it very fpeedily faded and died; poffibly injured by the folution of nitrous oxide in water.

Of two rows of peas juft appearing above ground; I watered one with folution of nitrous oxide in water, and the other with common water daily, for a fortnight. At the end of this time, I could perceive no difference in their growth, and afterwards they continued to grow equally faft.

I introduced through water into fix phials, one of which contained hydrogene, one oxygene, one common air, one hydrocarbonate, one carbonic acid, and one nitrous oxide, fix fimilar plants of mint, their roots being in contact with water and their leaves expofed to light.

3. The plant in carbonic acid began to fade in lefs than two days, and in four was dead. That in hydrogene died in lefs than five days; that in nitrous oxide did not fade much for the firft two days, but on the third, drooped very much, and was dead at the fame time as that in hydrogene. The plant in oxygene for the firft four days, looked flourifhing and was certainly of a finer green than before, gradually however, its leaves became fpotted with black and dropped off one by one, till at the end of ten days they had all difappeared. At this time the plant in common air looked fickly and yellow, whilft that in hydrocarbonate was greener and more flourifhing than ever.

I have detailed thefe experiments not on account of any important conclufions that may be drawn from them; but with a view of inducing others to repeat them, and to examine the changes effected in the gases. If it fhould be found by future experiments, that hydrocarbonate generally increafed vegetation, it would throw fome light upon the ufe of manures, containing putrefying animal and vegetable fubftances, from which this gas is perpetually evolved.

The chemiftry of vegetation though immediately connected with agriculture, the art on which we depend for fubfiftence, has been but little inveftigated. The difcoveries of Prieftley and Ingenhoufz, feem to prove that it is within the reach of our inftruments of experiment.

No. II.

APPROXIMATIONS

TO THE

Composition and Weight of the aëriform
COMBINATIONS of NITROGENE.

At temperature 55º, and atmofpheric preffure 30.

		100 Cubic In.	grains		Nitrogene	Oxygene
		Nitrogene	30.04			
		Oxygene	35.06			
Nitrogene	With Oxygene	Atmofpher.air	31.10	100 grains are compofed of	73.00	27.00
		Nitrous oxide	50.20		63.30	36.70
		Nitrous gas	34.26	weigh	44.05	55.95
		Nitric acid	76.00		29.50	70.50
					Nitrogene	Hydrogene
	Withhydrogene	Ammoniac	18.05		80.00	20.00

a. In Ref. 1ft. Div. IV. Sect. III. in the analysis of nitrous gas by pyrophorus, as no abforption took place when the refidual nitrogene was expofed to water, I inferred that if any carbonic acid was formed it was in quantity fo minute, as to be unworthy of notice. A few days ago, I compleatly decompofed a quantity of nitrous gas by pyrophorus, when the refidual nitrogene was expofed to folution of ftrontian, the fluid became flightly clouded ; but no perceptible abforption took place.

b. If there was the leaft probability in any of Dr. Girtanner's fpeculations on the compofition of Azote,* the experiments on the exhaufted capacity ‡ of the lungs in Ref. III. might be fuppofed inconclufive. But there appears to

* Annales de Chimie, 100; and Mr. Tilloch's Phil. Magazine. 24.

‡ I regret much that I could not procure Dr. Menzies's obfervations on Refpiration, while I was making the experiments on the capacity of the lungs : they would probably have faved me fome labor.

be no more reafon for fuppofing that hydrogene is converted into nitrogene by refpiration, than for fuppofing that it is converted into water, carbonic acid or oxygene; for all thefe products are evolved when that gas is refpired. From the comparifon of Exp. 1 with Exp. 3, Ref. iii. Div. ii. Sec. 4, it is almoft demonftrated that no afcertainable change is effected in hydrogene by refpiration. The experiment of the accurate Scheele in which hydrogene after being refpired thirty times in a bladder wholly loft its inflammability, may be eafily accounted for from its mixture with the refidual gafes of the lungs.

About a fortnight ago, I refpired, after forced voluntary exhauftion of my lungs, my nofe being accurately clofed, three quarts of hydrogene in a filk bag, at four intervals, for near five minutes. After this it was highly inflammable, and burnt with a greenifh white flame in contact with the atmofphere; but was not fo explofive as before.*

* If loofely combined carbon exifts in venous blood, hydrogene may probably diffolve a portion of it when

c. From what we have lately heard of the curious experiments of Mr. Volta and Mr. Carlifle, it is very probable that the converfion of nitrous gas into nitrous oxide when expofed to wetted zinc, copper and tin, in contact with mercury, as defcribed in Ref. I. Div. V. may in fome meafure depend on the action of the galvanic fluid. Whilft I was engaged in the experiments on this converfion, Dr. Beddoes * mentioned to me fome curious facts noticed by Humboldt and Ritter, relating to the oxydation of metals by the decompofition of water, which induced me

refpired and become flightly carbonated. At leaft there is as much probability in the fuppofition that carbon in loofe affinity may combine with hydrogene at 98° as that it may combine with oxygene.

* Dr. BEDDOES has fince favoured me with the following account of thefe facts.

" Mr. Humboldt (ueber die gereizte Fafer I. 473, 1797) quotes part of a letter from Dr. Afh, in which it is faid that *if two finely polifhed plates of homogeneous zinc be moistened and laid together, little effect follows—but if zinc and filver be tried in the fame way, the whole furface of the filver will be covered with oxydated zinc. Lead and quickfilver act as powerfully on each other, and fo do iron and*

to examine the phænomena with more attention than I should have otherwise done.—I recollect obferving that fome of the wetted zinc

copper.—Mr. Humboldt (p. 474) fays that, in repeating this
experiment, he faw air-bubbles afcend, which he fuppofes
to have been hydrogene gas from the decompofition of
water—When he placed zinc fimply on moift glafs, the
fame phænomena took place, but more flowly and later.
The quantity of oxyd of zinc upon the glafs alone was in
20 hours to that on the filver as one to three.

In a very ingenious but obfcurely written tract
by Mr. Ritter, entitled, *Evidence that the galvanic
action exifts in organic nature*, 8vo. Jena, 1800—The
author obferves, that the care of Dr. Afh and Mr. Humboldt that the metals fhould touch each other in as
many points as poffible was fuperfluous; even if we could
grant that two metallic plates might be made by polifhing,
to touch in a number of points. To fhew that it was fufficient if by touching in one point only they fhould form a
compleat galvanic circle, he dropped a fingle drop of diftilled
water upon the buft of a large filver coin. A piece of pure
zinc was placed with its one end on the edge of the coin,
while the other was fupported by a bit of glafs. The drop
of water was neither in contact with the glafs nor with the
point at which the metals touched. The mateiials were
left in this fituation for four hours at the temperature of
68°. On taking them apart, the water had become quite
milky and had half difappeared; and Mr. Ritter actually
feparated a quantity of white oxide that had been produced
in the experiment.

The pieces of metal were cleaned and laid together in

filings in nitrous gas on the fide of the jar not in contact with the furface of mercury, were very flowly oxydated. Whilft on the furface of the mercury where fmall globules of that fubftance were mingled with the filings of zinc, the decompofition went on much more rapidly;

the fame manner, only that now a piece of paper was put between the metals at their former point of contact. In four hours firft, and afterwards in ten, a faint ring of oxide only had been produced of which the quantity could not be eftimated, nor could it be feparated. In this cafe, the zinc had fcarce loft any thing of its fplendour; in the former it had been corroded. In many repetitions of the experiment, he found that far more oxide was formed when the metals touched, than when they were feparated to the flighteft diftance by an infolating body, even air.

On expofing thefe apparatufes with fomewhat more water to a confiderable heat for four minutes, the water in the interrupted circle continued quite clear, while that in the other had become milk-white.

The fame phænomena were prefented by other pairs of metals in a degree proportional to their galvanic activity; viz. by zinc and molybdæna, zinc and bifmuth, zinc and copper, as alfo with tin and filver, tin and molybdæna, and lead and filver. The experiment with tin was particularly decifive, for when in contact with no other metal it was fcarcely at all oxydated by water, though oxydation took place when tin was brought into contact with filver, poffibly through the medium of the moifture, a feries of galvanic circles were formed.

d. In Ref. II. Div. I. it is ftated, that nitrous oxide during its folution by common water, expels about $\frac{1}{16}$ of atmofpheric air the volume of the water being unity.

and both were connected at the other end by a drop of water—What therefore took place in Dr. Afh's experiment, arofe from an aggregation of galvanic circles of different forms.

By the foregoing experiments, concludes Mr. Ritter, which though capable of the moft various modifications, uniformly coincide in their main refult, it is, abundantly, proved that *galvanic circles can be formed of merely inorganic bodies, by whofe completion there is produced an action which ceafes when the circle is opened.* The manner in which this has been fhewn, proves alfo that *this action can effectuate fenfible modifications in organic bodies;* and the procefs by which thefe modifications have been effected, made it evident that they *were not confequences of a momentary action of the circle, but of an action that is kept up while the circle remains entire;* for the procefs which brought this action under the cognizance of the fenfes went on, while the circle was unbroken, and its figure not brought back to that of a line.

It is fcarce neceffary to obferve that the experiments here quoted, are far from being the only ones on which the above conclufions reft."

T. B.

From the delicate experiments of Dr. Pear-
son, on the paffage of the electric fpark through
water, it appears however probable, that much
more than $\frac{1}{16}$ of atmofpheric air is fometimes
held in folution by that fluid,* poffibly the
whole of the air is not expelled by nitrous oxide,
owing to fome unknown law of faturation by
which an equilibrium of affinity is produced,
forming a triple compound.

* Poffibly a ratio exists between the folubility of gafes
in water, and the folubility of water in gafes. It is probable from Mr. Wm. Henry's curious experiments on the
muriatic acid, that the abfolute quantity of water in *many*
gafes, may be afcertained by means of its decompofition by
the electric fpark.

No. IV.

,DESCRIPTION OF A

MERCURIAL AIR-HOLDER,

Suggefted by an infpection of Mr. Watt's Machine for containing Factitious Airs,

By WILLIAM CLAYFIELD.

Several modes of counteracting the preffure of a decreafing column of mercury having been, thought of in conjunction with Mr. W. Cox, the following was at laft adopted as the moft fimple and effectual.

Plate 1 Fig. 1, reprefents a fection of the machine, which confifts of a ftrong glafs cylinder A cemented to one of the fame kind B, fitted to the folid block C, into which the glafs tube D is cemented for conveying air into the moveable receiver E.

The brafs axis F, Fig. 2, having a double bearing at *a, a,* is terminated at one end by the wheel G, the circumference of which is equal to the depth of the receiver, fo that it may be drawn to the furface of the mercury by the cord *b* in one revolution; to the other end is fitted the wheel H, over which the balance cord *c* runs in an oppofite direction in the fpiral groove *e*, a front view of the wheel H is fhewn at Fig. 3.

Having loaded the receiver with the weight I, fomething heavier than may be neceffary to force it through the mercury, it is balanced by the fmall weight K, which hangs from that part of the fpiral where the radius is equal to that of the wheel G, from this point the radius of the fpiral muft be increafed in fuch proportion, that in every part of its circuit, the weight K may be an exact counterpoife to the air-holder. In this way, fo little friction will be produced, that merely plunging the lower orifice of the tube D under mercury contained in the fmall veffel L, will be fufficient to overcome every refiftance, and to force the gas difcharged from the beak of a retort into the receiver, where whatever may be its quantity, it will be fubjected to a preffure exactly correfponding to that of the atmofphere. The edge of the wheel H being graduated, the balance cord *c* may be made to indicate its volume.

Should it at any time be neceffary to reduce

barometer, it may eafily be done by graduating
the lower end of the tube D, and adding to the
weights I or K, as may be found neceffary ; the
furface of the mercury in the tube pointing out
the increafe or diminution.

The concavity at the top of the internal
cylinder is intended to contain any liquid it
may be thought proper to expofe to the action
of the gas.

The upper orifice f, with its ground-ftopper,
is particularly ufeful in conveying air from the
retort g, with its curved neck, into the receiver,
without its paffing through the tube D. In all
cafes where a rapid extrication of gas is expected
the retort g, fhould be firmly luted to the orifice, and the weight I, removed from the top
of the receiver, this by diminifhing the preffure, will admit the gas to expand freely in the
air-holder at the inftant of its formation, and
prevent an explofion of the veffels. The fame
caution muft be obferved whenever any inflammation of gas is produced by the electric fpark.

The air may be readily transferred through
water or even mercury by the tube h, Fig. 1.

To prevent an abforption of mercury in cafe of a condenfation taking place., in, the retort made ufe of for. generating air; Mr. Davy has applied the ftop-cock *i*, to which the neck is firmly luted. · This ftop-cock is likewife of great fervice in faturating water with acid or alkaline gafes, which may be effected by luting one end of the tube *k*.to the ftop-cock,.and plunging the other into the flu.d in the fmall veffel *l*, cemented at top, and terminating in the bent. funnel *m*—the tube *h* having been previoufly removed, and the lower orifice of the tube *D* either funk to a confiderable depth in mercury, or clofed with a ground ftopper. The bend of the funnel *m*, may be accurately clofed by the introduction of a few lines of mercury.

The application of the ftop-cock *n*, has enabled Mr. Davy to perform fome experiments on refpiration with confiderable accuracy.

Note. This apparatus was firft defcribed in the third part of Dr. Beddoes's Confiderations; its relation to Mr. Davy's experiments with the improvements it has lately received, may probably be deemed fufficient to excufe the re-printing it.—The weight I. Fig. 2, having been omitted in the plate, the reader muft fupply the deficiency,

W. C.

PROPOSAL

THE PRESERVATION

OF

ACCIDENTAL OBSERVATIONS

IN

MEDICINE.

IN times beyond the reach of hiftory, the medicinal application of fubftances could have arifen from no other fource than accident. Among articles of the materia medica of known origin, we are indebted to accident for fome of the moft precious.

Accident is every day prefenting to different individuals the fpectacle of phænomena, arifing from uncommon quantities of drugs on the one hand, and on the other, from uncommon conditions of the fyftem, where ordinary powers only have been knowingly or recently applied. What is faid of drugs may be extended to natural agents and mental affections.

From converfation with a variety both of medical practitione, and unprofeffional obfervers, the author of this propofal is perfuaded that fuch authentic occurrences only, as have prefented themfelves to perfons now living would, if they could be brought together, compofe a body of fact, fo inftructive to the philofopher, and ufeful to the phyfician, that he defpairs of finding a term worthy to characterize it.

In·fome cafes, the influence of unfufpeâed powers would be deteâed. In others, refources available to the purpofe of reftoring health in , defperate fituations would be direâly prefented,' or could be deteâed by a fhort and eafy procefs of reafoning. Some anomalous obfervations, by fhewing the abfence or agency of contefted caufes, would perform the office of *experimenta crucis*—Unufual affeâions occur of which an exaâ account would be among the means of removing from phyfic its opprobrious uncertainty: for this uncertainty frequently depends upon our inability to diftinguifh the fubtler differences in, cafes which refemble each other in their groffer features.

No ftriking faâ can be accurately ftated, in conjunâion with its antecedent and concomitant circumftances, without improving our acquaintance with human nature. Our. acquifitions in this moft important branch of knowledge, may be compared to a number of broken feries, of which we have not always more than one or two members. But every new acceffion bids fair to fill up fome deficiency; and a large fupply would contribute towards conneâing feries apparently independent, and working up the whole into one grand all-comprehending chain.

There are complaints, and thofe by far too frequent, where no known procefs has a claim to the title of *remedial*. Here the whole chance of prefervation depends on the phyfician's capacity for bringing together faâs that have heretofore ftood remote. But no power of combination can avail where there are no ideas to combine.

Every new obfervation therefore, may be, confidered as a ftandard trunk, fending forth analogies as fo many branches crowned with bloffoms, fome of which cannot fail to be fucceeded by falutary fruits. And were it not abfurd to extend the illuftration of fo plain a point, it might be added, that when by the continual interpofition of new trunks, the branches are brought near together, the produce of each, will be ennobled by the action of their refpective principles, of fecundation.

Whenever the author has been able to obtain certain information concerning any unufual appearance in animal nature, it has been his cuftom to preferve it; and among his papers he has memorandums which prove that to our prefent circumfcribed ideas concerning the dofe of medicines may be fometimes imputed failures in practice; that certain figns are not to be taken in, the received fignification; and that many meafures are adopted or omitted to the detriment of invalids, becaufe it is affumed that circumftances are neceffarily connected, which may exift feparately, or that one given natural operation is inconfiftent with another, to which it may really be fynchronous or next in order.

Affiduous obfervation of the daily ftates of the human microcofm will be the unfailing confequence of attention to its ftriking phænomena. Such is the progrefs of curiofity. Such the origin of all the fciences. The more uniformly clear the fky under which they tended their flocks, the lefs likely were the fhepherds of Chaldæa, to found the fcience of the ftars. And however the difpofition to ftudy aftronomy might have been ftrengthened by the coincidence between the heliacal rifing of Sirius and the overflowing of the Nile, it muft, I conceive, have been awakened by the afpect of meteors and eclipfes.

Whatever minute and authentic information this imperfect ftatement may produce, as foon as it fhall amount to a certain mafs, the author will prefent it to the public arranged. He flatters himfelf that no correfpondent will eke out by fuppofition the defect of genuine obfervation, without clearly diftinguifhing the one from the other. He ftill more confidently hopes that none will be inftigated by this advertifement to exercife his invention in the manner of Pfalmanafar and Chatterton. Whether any literary forgery can be innocent is queftioned—but a forged medical report is a drawn dagger which the arm of a credulous phyfician may any day plunge into the heart of his defencelefs patient. The author has heard fome inconfiderate wits avow, that they have tranfmitted to the venders of quack medicines imaginary cures, attefted by fictitious fignatures; and it is not without apprehenfion from the propenfity of men to difplay ingenuity and to relate wonders that he announces the prefent defign. But he fhall be on his guard, and hopes to baffle attempts at impofition.

THOMAS BEDDOES.

Rodney-Place, Clifton, June 1800.

END.

ERRATA.

Page 19 line 15 for *is* read *are*

— 35 — 7 — for *principle* read *principles*

— 42 — 11 — for *take* read *takes*

— 68 Table 5 — for 5,88 read 15,88

— 94 — 4 — for $1\frac{1}{12}$ read $\frac{1}{12}$

— 95 — 4 — for 37 read 30,7

— 96 — 3 — for 38 read $\frac{1}{38}$

— 105 — 9 — for *exactitude* read *exactnefs*

— 129 — 21 — for 41 read 4,1

— 132 — 4 — for *into* read *in*

— 143 — 13 — for 25 read .25

— 186 — 15 — for *by* read *from*

— 208 laft line — for *abftracted* read *attracted*

— 238 — 5 — for *gas* read *oxide*

— 259 — 4 — for 12 read 2

— 283 — 4 — for *potafh* read *iron*

— 315 — 14 — dele *in*

— 409 — 15 — for *refpiration* read *expiration*

— 464 — 10 — for *latter end* read *end*

— 543 — 3 for *exhalation* read *inhalation*.

A few literal errors are left to the reader's correction.

N. B. The term ignited is fometimes ufed to fignify any temperature equal to or above a red heat, whether applied to folids, fluids, or aëriform fubftances.

The reafons for the ufe of the terms nitrogene and nitrous oxide, are given in Mr. Nicholfon's Journal for January.

Speedily will be Published,

OBSERVATIONS on the External and Internal Ufe of

NITROUS ACID.

Demonftrating its PERMANENT EFFICACY in

VENEREAL COMPLAINTS;

And extending its ufe to other dangerous and painful
Difeafes.

COMMUNICATED

By various Practitioners in EUROPE and ASIA,

TO

THOMAS BEDDOES, M. D.

———————

Of the Publisher may be had, price 1s. 6d.

NOTICE of OBSERVATIONS

AT THE PNEUMATIC INSTITUTION,

By THOMAS BEDDOES, M. D.

This Notice contains fome trials of nitrous oxide by healthy perfons,
not in the prefent work, and fome cafes of palfy fuccefsfully treated
by that gas.

Printed by Biggs and Cottle, St. Augustine's Back.